THE DIARY OF A TEA PLANTER

THE DIARY OF A TEA PLANTER

F.A. Hetherington

The Book Guild Ltd
Sussex, England

The Book Guild Ltd
25 High Street,
Lewes, Sussex

First published 1994
© F.A. Hetherington

Set in Baskerville

Typesetting by Wordset
Hassocks, West Sussex

Printed in Great Britain by
Antony Rowe Ltd
Chippenham, Wiltshire.

A catalogue record for this book is
available from the British Library

ISBN 0 86332 935 7

CONTENTS

Preface   vii
Tea in Assam   viii

1   Journey to India, 1900 *(First Diary)*   1
2   Monabari Tea Estate   10
3   Light Horse Camp at Dibrugarh   44
4   In Charge at Deckajuli   96
5   Back at Monabari *(Third Diary)*   101
6   Going on Home Leave   104
7   My Second Tour of India   121
8   Marriage in Calcutta   171
9   Ruthna Tea Estate   203
10 Return Home   238

    Appendix: Work Rota   246
    Diagramatic Map of Assam and Sylhet   248
    Glossary   249
    Letter of Appointment   251

LIST OF PHOTOGRAPHS

Burra Bungalow, polo ponies and buggy.
F. Troop Assam Valley Light Horse on parade.
Pruning tea bushes.
Planters and wives. Writer of Diary cross-legged on right.
F. Troop AVLH.

PREFACE

The author of the diary was born in Wetheringset Rectory in 1878 where his father was the curate. He was christened there and one of his godfathers was Sir Edward Kerrison of Brome Hall.

When he was four the family moved to West Bradenham when his father was appointed vicar. William Haggard, the father of Rider, was the squire at that time.

After prep school he was educated at Hailebury college and then trained as an engineer at Harland and Wolf in Belfast.

When fit again after being invalided from India he tried his hand at farming but it was not a success. He then went to Canada but the climate did not suit him and he returned to England. He served with the Admiralty during the war and then retired to Staines where he died in 1939.

These abstracts were made by his son, H. A. Hetherington, who after qualifying as a civil engineer went to India in 1933 by P & O steamer and train, in much the same way as his father. He worked on the provision of water suplies to tea estates and jute mills and travelled on river steamers in Assam as described in the diary. He visited India again in 1970 as consultant to WHO to report on village water supplies and found in many ways a different country, though the country folk had hardly changed and were the same pleasant people.

TEA IN ASSAM

Tea was first reported growing in Assam in 1823 but thought not to be the same species as the plant from which the Chinese made tea. Later, at the end of 1835, it was proved that the tea of Assam was the true tea of commerce, but planters found that the best plants were a hybrid of Chinese and Assam plants. In 1840 the Assam company established a factory at Dibrugarh and in 1859 had 4,000 acres under cultivation. In 1872 about 27,000 acres were planted with tea and in 1901 approximately 72 million pounds of tea were produced in the Brahamaputra valley.

The system of manufacture in 1904 was to lay out freshly plucked leaf thinly on trays in order that it might wither and become soft and easily rolled. When withering was complete the leaf was rolled to exclude the sap. As soon as the leaf cells were broken up the leaf was taken out of the rollers and allowed to stand until fermentation had set in. The leaf was then placed in a drier and fired at a fairly high temperature which turned the colour of the leaf to nearly black. It was then sorted, sieved and packed in tea-chests.

In 1847 a government steamer service was established but the boats did not go beyond Gauhati. By 1906, however, a regular service was available, the upward journey from Goalundo to Dibrugarh taking less than a week. The steamer service on the Brahamaputra and Surma rivers would probably not have been developed but for the trade in tea. There was insufficient labour available in Assam to work on the tea-gardens so labour had to be imported generally from Bengal.

1

JOURNEY TO INDIA, 1900

(First diary)

Wed. Aug. 15th, 1900

Grand Hotel, Charing Cross. Breakfast, 7.45 a.m. Then went to the station and registered my luggage. McLeod and Co. had given me a first class ticket, for which Allah be praised. Left at 9 a.m. Father and Mother rather cut up. Strange that I don't feel that way myself.

Only one other passenger, a young chap going to Australia. Offered him a cigar. He only took cigarettes; used a silver case. The Kentish hop fields looking lovely. The whole journey new and interesting. Just before Dover lit the last of the Corona cigars; my last smoke in the old country. Went on board and got my luggage together. Just as we ran alongside Calais pier I went aft and waved my cap to the Old Country. Wonder if I shall ever see it again?

At Calais got my luggage passed with the help of an interpreter, who very pointedly asked for a tip. Then took a good seat in the diner for my first meal abroad: omelette, hash, cold tongue, cheese and fruit. My vis-à-vis seemed to travel for some firm. We sat together and smoked afterwards.

French scenery looked much the same as the English. The train was splendid, about fifty miles an hour. Just as we reached Paris my companion pointed out the Eiffel Tower. It is a marvellous-looking structure. At the Gare du Nord an interpreter put me and two more into a cab and we crossed to the PLM, a fine-looking station. Had

1

some trouble finding an English-speaking porter but got him at last and put my coat away. Then went on a walk, crossing the Seine. Watched some river steamers covered with advertisements for chocolate Menier – how familiar it seems! – and returned to the station for dinner. Went out again and bought a wooden cigar holder as a memento for fifteen cents. Left Paris at 8.40 p.m. Lovely moonlit night.

Thurs. Aug. 15th

Reached Lyons at 4 a.m., Avignon at 7 a.m., where we got coffee and Marseilles at 9 a.m. I gave my coat to a King's man who took it and directed me to the Hotel de Russie for breakfast. Here I found an English-speaking waiter. Washed and breakfasted and then started for the boat. The little tables outside the wine shops reminded me of the scene in *La Bohème*. After reaching the quay a 'working man' piloted me to the *Himalaya*. The gangway was surrounded by fruit and chair sellers. Found my cabin steward – Abbott – a little yellow-haired fellow; no luggage had arrived. Wrote home, then dinner; a nice seat against the saloon side. Took my letters to King's and inquired about my coat. Returned on board and found my luggage and coat had arrived, for which I paid.

Left about 4 p.m. Watched the warping out and the coolies at work. Passed the Château d'If, then went to tea and took stock of the passengers. A very mixed lot. After tea walked the deck till 9 p.m., then turned in. The cabin very small and mostly full of my companion's bags, marked 'J. Rosenthal, War Correspondent'. My thermometer showed 80°, but I slept well.

Fri. Aug. 17th

Steward called me at 7.30 a.m. Rosenthal greeted me with

the top of the morning. Medium height, dark-haired, moustache and pince-nez; also swears like a trooper. Told me he was a cinematograph man for the *Illustrated London News*. Has just returned from Africa and is going to China. After breakfast went on deck, where I found a chair, smoked and then bathed. Passed through the Strait of Bonifacio. Both islands misty, also interesting; Corsica, on account of Napoleon. In the afternoon got in conversation with a gentleman called Yule who turns out to be a brother to General Yule of Retreat to Ladysmith fame. He told me he has a jute factory near Calcutta; was a Rugby boy while the general is an old Haileyburian. He is taking his son out to get a job on the railways. Impromptu concert in the evening; very passable but could sing as well myself. At Yule junior's advice had my bed on deck and found it very pleasant. Saw a man who looks like Shucks Henderson of Thomason.

Sat. Aug. 18th

Woke at 6 a.m. when the decks were cleared for washing. Went below and dozed again. At breakfast time we passed Stromboli − not, unfortunately, in a state of eruption, though a cloud of smoke was hanging over the crater. The island is a magnificent crag with vineyards looking cool and green covering the lower slopes and a little town on the side away from the crater. About midday the baggage was brought up and I got out my flannels and changed. Had a yarn with Henderson, who is the man right enough. He says he is in the Madras Lancers.

During the morning we passed through the Straits of Messina, the shore all steep cliffs and very rugged. Every hour the temperature gets hotter. In the evening M. Neufeld, the man who had been the Khalifa's prisoner, gave us a lecture illustrated with views and wonderfully interesting. He told us of his expedition after gum, which resulted in his capture, sentence of death, flogging − 150 lashes – experiments in saltpetre, marriage to an Abyssinian, witnessing of the execution of other prisoners and final release by Kitchener. He speaks English very well, is rather a nice-looking man above medium height with a long brown beard. Slept on deck. Warmer than ever.

Sun. Aug. 19th

Service in the first saloon which I did not attend. Rather rough after dinner and I was a little sick.

Mon. Aug. 20th

Played off my whist match and got beaten. Attempted dancing in the evening but not very successful. I found that it was not so easy waltzing on deck as I had thought and gave it up as the girls were dull. Had an interesting talk with Mr Douglas, the missionary. Found his people knew the Mills of Great Cressingham.

Tues. Aug. 21st

Reached Port Said about 11 a.m., and we commenced coaling. Noticed statue of de Lesseps on the breakwater. Several ships in port, including a big three-funnel, three-mast, fiddle-bowed Russian trooper. Looked round the town, was touted for can-can, etc. Returned to ship and talked to Douglas and Pryke, who is going to Borneo for railway construction.

Wed. Aug. 22nd

Orient Line *Austral* passed through for London. We left about 11 a.m. Saw camels in their native condition for the first time. Reached Ismailia at 7 p.m. and a few passengers landed. Searchlights on the Canal looked very pretty; all ships obliged to use them.

Thurs. Aug. 23rd

Passed through the Gulf of Suez, the hills on both sides covered with a pink mist. Concert in the evening, followed by a magnificent raid on the hurricane deck, held by the officers' squad. Pryke and I attacked by way of the rigging; final successful storm led by Yule senior.

Fri. Aug. 24th

The anniversary of the *Oceanic's* trial. How well I remember it, leaving Belfast quay at 10.45 a.m., Miss Chattie and the children watching us off and the lovely moonlit journey home in the tug!

Sat. Aug. 25th

We had a big dance but I didn't join in, as it was too hot.

Sun. Aug. 26th

Paid my stewards and packed. We reached Aden about 3.30 p.m. and lay at a good distance from the land. The ship was soon surrounded by boats, some with boys shouting, 'Have a dive!', others with skin sellers, etc. We left the *Himalaya* at 5 p.m., in a tug, reaching the *Oriental* in about ten minutes. I got a cabin to myself but it was too stuffy to sleep in. We left about 8.30 p.m., an intensely hot wind blowing. Aden appears a most desolate place, with nothing to recommend it.

Mon. Aug. 27th

Joined the ship's library and read all day. Not feeling very well, so turned in early.

Tues. Aug. 28th

Very rough. Didn't feel at all well.

Wed. Aug. 29th

Headache very bad, so turned in after breakfast. Saw the doctor in the afternoon.

Sat. Sept. 1st

Felt better. Got up at 5.30 a.m. and went on deck to get my first view of India, but on account of clouds it was rather disappointing. Anchored at 7 a.m. and then went off in launch to Ballard pier. Luggage passed without difficulty and drove to Victoria Station, which has the finest entrance hall of any that I know, with marble pillars and painted windows. Very comfortable carriage arranged for four but only Captain Oswald and I in it. Slept in pyjamas and greatcoat.

Sun. Sept. 2nd

Train passed through good tiger country but we weren't lucky enough to see any. About 1 p.m. the train whistled and pulled up and on getting out, found that a coolie had been killed. Though it was the first corpse that I had seen, somehow I felt an oriental indifference. Two coolies picked up the body and cast it beside the track. The restaurant man remarked, 'They killed a goat there the week before', and we moved on.

Mon. Sept. 3rd

Woke about 5 a.m. to find we had reached Howrah. Said goodbye to Oswald and drove to the Great Eastern Hotel. Turned in till 8.30 a.m., then breakfasted and drove to McLeods. Heard I was to go to Monabari, as the last engineer is being transferred for incompetence. After dinner drove to Sealdah Station and got a good carriage. Left at 10 p.m. and turned in immediately.

Tues. Sept. 4th

Wakened about 4 a.m. by a man feeling my pulse as a plague examination. Twenty minutes later we arrived at Goalundo Ghat and I boarded the river steamer, the *Duffla*. Had a very comfortable cabin on the little upper deck with two doors, one opening to the deck and the other to the saloon. Nothing to do on board but eat, read and sleep, lulled occasionally by the monotonous singsong voice of the leadsman: *sari-ek-bahn* (a fathom and a half) or *ek-bahn-ek-hat* (a fathom and a quarter). Meals were *chota hazri*, 9 a.m., *burra hazri*, about 12.30 p.m., tea 4-5 p.m., and *khana*, 7-8 p.m. Made the acquaintance of guava jelly and tiparee jam, the latter a kind of gooseberry and very delicious. We stopped at several river stations, or ghats, where we moored to barges, or flats.

Wed. Sept. 5th

Stopped at Fulcharighat to pick up a Captain Walling of the 5th Bengal Light Infantry. Got on well with him. A fine sunset on the water.

Thurs. Sept. 6th

Walling pointed out places for crocodiles, or *muggars* as they are called here. The river narrower and more interesting. Reached Gauhati about 8 p.m. The river narrow here between hills. Walling said it is a lovely spot to arrive at on a fine day.

Fri. Sept. 7th

Captain Walling pointed out some crocodiles, or *muggars*, lying on sandbanks. We kept close to the bank with the ground behind swampy and covered with tall grass. Got my first view of the hills, but it was a poor one, owing to tremendous rain clouds. A fine sunset, with a lurid glow on the mass of storm cloud. Reached Tezpur at 6.45 p.m. Glad I didn't have to land, as it was dark.

Sat. Sept. 8th

Walling pointed out two snowclad peaks where the clouds were broken. So this was my first view of the everlasting snows. We reached Behalimukh at 11 a.m. Landed and found a pony and syce with a letter from Godwin to tell me to send my luggage by boat and ride a distance of about four miles. Unable to mount, as the pony kicked and bucked, so decided to walk. Country flat and marshy, covered with reeds. Saw some splendid butterflies. Arrived at a river and got a punt sent across; a coolie carried me to it. Then, after a short distance, saw the coolie sheds and, coming closer, saw my first tea garden.

The plants are little dark green bushes, about three feet high. Disturbed a small snake, which glided away. I have nearly got over my old loathing of them. Passed coolie lines and the tea house and, on approaching the bungalow, heard a voice calling me and found on the verandah my future boss, Colin Dunlop; fairly tall, darkish hair and auburn moustached, and the senior assistant, Godwin; slightly taller, very dark with black moustache. Godwin lent me a shirt to wear as mine was soaked with sweat. Had *burra hazri* and then was shown round the factory and saw Griffith, the outgoing engineer, a rather slack-looking customer. Had dinner and then smoked on the verandah while the fireflies made darts like electric flashes. The moon threw a silver light on the compound and the booming of tomtoms came from the village.

2

MONABARI TEA ESTATE

Sun. Sept. 9th

Godwin woke me about 5 a.m. to say a paragon had broken down and Griffith was working on it. The tea house works on Sunday but not the pluckers. Went down to assist and sweated worse than ever before in my life. Got the paragon working again in the afternoon. The bungalow consists of sitting room, dining room and two bedrooms. There is a piano, thank goodness, though a poor one. I was to sleep in Godwin's room and turned in about 11 p.m., pleasantly tired.

Mon. Sept. 10th

Another small breakdown in thee paragon, which I repaired. Griffith left. Had my first ride in the afternoon, Godwin coaching. Dunlop seems to have a good opinion of Haileybury.

Tues. Sept. 11th

Am beginning to learn a little Hindustani, and helped at the midday weighing. No ride, as Hannen came over to tennis. Slept in the chota bungalow for the first time. It is a palatial establishment of two rooms and a bathroom built of bamboo, covered with dry mud and lighted with a stable lantern. My boy is called Nando Lall and does me very well.

Wed. Sept. 12th

Dr Smith came over to breakfast. Like the look of him. Rode in the afternoon and got my first tumble, which did not hurt and gave me more confidence. Find I can do very well on non-alcoholic drinks. We make our own soda water and ginger ale.

Thurs. Sept. 13th

Rode in early morning. Worked till breakfast at weighing boxes. Hannen came over to tennis. It is pleasant having six or seven coolies to pick up balls.

Fri. Sept. 14th

Rode in the compound as usual. Tea work weighing and marking boxes. What a relief we don't dress for dinner; so no more stiff shirts!

Tues. Sept. 18th

Watched a large rat snake killed near the stables. Mr and Mrs Swinley came over but didn't stay for dinner.

Wed. Sept. 19th

Dunlop drove me over to the Swinleys at Mijica. On the way some coolies with leaf carts told us they had been followed by a tiger. The Swinleys' bungalow is simply

11

beautiful and the garden well kept. Was introduced to their engineer, Lecky, and find he is an old Haileyburian. Had a long chat and went round their factory. They have two siroccos and a box-packing machine, also a sawmill. A tramway brings in their leaf and a branch line runs into the forest for wood.

Sat. Sept. 22nd

Dunlop went to polo. I have taken over Sirung and went out on her twice. Stoppage of paragon.

Sun. Sept. 23rd

Fixed up paragon. Short day's work; all over at 6 p.m. Hannen and Garlant came over for tennis.

Tues. Sept. 25th

Nothing worth noting, except the capture of a young cobra in the cookhouse.

Sat. Sept. 29th

Rode to Mijica. Called on the Swinleys and then the Smiths; my longest attempt, twelve miles. The feature of the Smiths' bungalow is Kruger, a tailless monkey called by the coolies 'the Jungli Man', some paying *pice* or quarter of an anna to see it. Watched polo on the way back.

Mon. Oct. 1st

The first day that felt cool, or rather, a suspicion of coolness.

Sat. Oct. 5th

Spent the night at Lecky's bungalow. Felt sick and chilly at midday. Lecky took my temperature – 104° – so I went to bed. Dr Smith came and gave me a rub down. My first dose of fever.

Sat. Oct. 13th

Back at work, feeling A1. Packed boxes, the last great pile of tea. Total boxes now about 950.

Thurs. Oct. 18th

Same endless box work. No chance for a ride or even to write a letter.

Sat. Oct. 20th

Sorted 108 boxes of Pekoe Souchong and repaired a rolling machine. Godwin went to polo.

Sat. Oct. 27th

In the evening went to a native performance given by the Jemadar, acted by professionals. We sat at a table with whisky and cakes. The play was about an ancient king who was dressed to appear ten feet high and the masks were splendid. We came away about midnight but the proceedings went on nearly all night.

Sun. Nov. 4th

Repaired the old paragon. Chichester came over to breakfast and we had four good sets of tennis, Godwin and I winning three.

Tues. Nov. 6th

Went out with Godwin and learned a little garden work. Saw to the cutting of a jungle line to prevent fire.

Sun. Nov. 11th

Two days' leaf worked off in the tea house. Started about 5 a.m. No one came to tennis and we had a little by ourselves.

Thurs. Nov. 22nd

Hannen arrived at 10.30 a.m. and he and Dunlop left for the polo match at Nowgong. Godwin went to Partabghur for drill and polo.

Sun. Nov. 25th

Packed 129 chests of tea. No tennis. Thank goodness one day's packing will see us out.

Mon. Nov. 26th

Dunlop and Hannen came back from Nowgong, having won by five goals to nil.

Fri. Nov. 30th

Rode to new ghat to inquire about cement, then to the Smiths at Behali for breakfast. Went round his lines, which are mud-built with corrugated roofs and very clean. Returned by way of Kettla, where I found Noble and Swinley, finally returning home at 5.45 p.m. Found Dunlop and Godwin had gone to Mijica for the night.

Sat. Dec. 1st

Dunlop came back early. I rode down to the brickfield and on returning felt sick again, temperature 104.2°. Better in the evening.

Mon. Dec. 3rd

Started clearing ground for the new leaf house and sorting out steelwork.

Sun. Dec. 9th

My first day with no tea house work. Walked round with Dunlop and Godwin to try bushes for pruning. Then for my first shoot. Only fired one shot – missed. Godwin got two partridges.

Mon. Dec. 17th

Last day's packing; over 13,000 lbs. Sixty seven maunds over the estimate and satisfactory on the whole. Total output 4367 maunds. Engine stopped and all belting removed.

Tues. Dec. 25th

My first Christmas away from home. Read Rudyard Kipling's *Christmas in India*. Left for Mijica about 11 a.m. and put up at Lecky's. Breakfast and then tennis. Had the luck to win all my sets. When it grew dark we had a first class tree on the verandah and all got presents at the hands of Bubbles, mine being an ashtray and a little cup. The tree looked very natural with its candles and flags. Then we got the sweepstake tickets ready, had dinner with the toast of absent friends and finished with music.

Mon. Dec. 31st

How well I remember last year, ariving in Belfast and getting no thanks from the children for my presents! Browning says, 'They have stabbed me with ingratitude. Well, in a thousand years' time we shall forget all the things that trouble us now.' Then I went in the evening to St George's with Chattie. Band, the curate, preached and I made good resolves – only to be broken. Then we came back and saw the New Year in. How it all comes back! But here I worked at the tea house and Godwin persuaded me to go to Hannen's, though I felt disinclined. I went, though, and Garlant and Lecky turned up and we had a great evening, the six of us singing 'Auld Land Syne' at 12 o'clock. Then Dunlop fired three shots over the grave of the departed year.

Tues. Jan. 1st 1901

There was moonlight and Godwin and I danced the Washington Post on the grass. Then we turned in for a few hours. Got up about 8 a.m. and spent the morning in the forest; then went to the polo ground.

16

Thurs. Jan. 24th

Weather a little warmer but am feeling a bit feverish. Brown and Swinley arrived for breakfast. Godwin went to meet Miss Chamney; rather liked her when we met. She brought news of the Queen's death. How strange it seems to think there will be no more 'God Save the Queen'! *Requiescat in Pace*, she deserves to.

Sun. Jan. 27th

Godwin shot over Kationabari; bagged one partridge and a couple of duck. I took the dogs and put up about ten birds, wounded one which went into the *bheel*, and then knocked one down – my very first – and I have never felt so pleased. Civilisation at the best is a very thin veneer. Noble came for breakfast and we had tennis as usual.

Sat. Feb. 2nd

No polo, owing to its being the Queen's funeral. A service was held at Tezpur.

Sat. Feb. 9th

I left at 3.15 p.m. for Borpukhri, where I stayed a little time, arriving at the Macraes first. All the down country men turned up and I was introduced to Davidson of Diharoi, Malcolm of Partabghur and Crutwell of Behali. There was a first rate dinner and then some theatricals, which were very good. Mrs Macrae, Edwards, Lillifant and Young took part. We sang 'God Save the King' for the first time and finished about 2 a.m.

17

Tues. Feb. 12th

We are getting back to the heat; one day has been over 80°.

Tues. Feb. 19th

I finished putting up the bungalow roof principles. Hot; started wearing white clothes.

Mon. Feb. 25th

Got steam up for the first time this year. It seemed quite strange to see smoke coming out of the chimney. Only put 20 lb of steam on as the manhole door leaked.

Sun. Mar. 10th

Went over to Kettla to bring Noble back, but neither he nor Dey will come, fearing a coolie row in their absence.

Wed. Mar. 20th

Got news that Hannen's flat had arrived and as Sirung was out went down on the bike. The captain and his mate were landing a box containing the engine, which weighed over two tons, with a hawser which broke in rather an alarming manner, while I was standing by. Went back to breakfast and sent my pony down for Captain Elder, but she wouldn't let him mount, so he walked up. I showed him round the place and afterwards we dined. Dunlop and Godwin escorted him back.

Fri. Mar. 22nd

Finished the leaf house. An elephant brought Hannen's engine through and Dunlop photographed it.

Sun. Mar. 24th

Dey, Noble and Crutwell came to breakfast. We used the punkah for the first time. Crutwell stayed the night and Godwin and I were initiated into bridge.

Mon. Mar. 25th

First day's plucking for the year, resulting in five maunds.

Tues. Mar. 26th

Started manufacture with one rolling table and a sirocco.

Thurs. Mar. 28th

Got some permanganate of potash for my feet from the doctor. Hope it will take off a little of the soreness. Felt most uncomfortable; nervous about some unknown event happening.

Sun. Apr. 7th

Easter Day. Two years ago it was our last Sunday at Bradenham. I read the lessons in the morning and Ed in the afternoon, and Dad preached his farewell sermon, to the accompaniment of sobs. Seventeen years is a good slice of a man's life. Here I went round the lines with Dunlop and out shooting, but no bag. Garlant, Lecky, Hannen and Noble came to breakfast.

Note in the Diary, dated 1931:

While talking to Dunlop Captain Elder alluded to me as a 'Jimmy-Come-Lately'. The name stuck to me and I became proud of it when I managed Deckajuli after only two years and nine months in tea, and I always signed myself 'Jimmy'.

Mon. Apr. 8th

Last year took Miss Chattie to Newcastle. How the Dad would have fumed if he had known! But all the boys were away and I was thrown on my beam ends and not unpleasant ones either. It was a perfect day and we walked up through the demesne to the summer house. The wind was whispering amid the pines, in contrast to the babble of the little mountain torrent, which was spanned by a rustic bridge (part of Browning's poem, *By the Fireside*, just describes it) and through a break in the trees. What a perfect view there was! The glistening blue of the bay with the lighthouse showing up white between Killough and Ardglass. The little town below looking so diminutive with a few specks out on the golf links conspicuous against the sand, and further on the salt water lagoon, known as Dumdrum Harbour; while fading in the distance lay the Ards peninsular. Here, I worked.

Wed. Apr. 11th

Manufactured again, but only three maunds. No rain, everything backward.

20

Sat. Apr. 13th

Rode down to the coal ghat with Dunlop; Godwin has fever. To polo and got very wet on the way back. Rain seems to be setting in at last.

Thurs. Apr. 18th

Started bungalow well. The ring arrived for it.

Sun. Apr. 28th

Bottled fifteen dozen whisky. Dunlop gets his sent up by the cask.

Wed. May 1st

Thirty maunds of leaf manufactured.

Sat. May 4th

Hospital well finished and water reached in the bungalow one. Made the two rooms of my basha into one.

Mon. May 20th

Rode to Bargang for breakfast to play Dr Smith in the billiard tournament. Had to use the ferry. Met Dunlop and Hannen there and walked up to the bungalow. Fever came on immediately so I couldn't play my match. In bed till evening and violently sick. River sank and syce crossed.

Tues. May 21st

Not much better. Could eat nothing. My boy came over with change of clothes.

Wed. May 22nd

Sent syce and pony back, feeling rather weak. Dey went to polo and made arrangements with Dunlop to fetch me back.

Sat. May 25th

Went on my work round about 5 p.m., when it got cool. Feeling a little better.

Sun. June 16th

What Crutwell calls a 'Europe morning'; in other words we took things easy. Dunlop left at 11 a.m. for local board meeting at Tezpur. Hottest day yet – 98°. Crutwell arrived 4.30 p.m. and we had indifferent tennis. C. stayed the night and we played bridge.

Mon. June 17th

Drove Crutwell to the Upper Bargang ford with Sirung. Got beaten by Dr Smith at billiards. Rode Lassie back. She came full speed most of the way and we were both fairly tired at the end.

Tues. June 18th

Very wet day. Killed a rat snake, five foot eight inches, and extracted young mynah.

Sun. June 23rd

Rode over to Behali and put up at Crutwell's bungalow. Glass, Dey and Noble came over shortly after and we adjourned to the burra bungalow for the christening of Gertrude Jane. The service rather a farce as the Smiths are Presbyterians.

Tues. June 25th

Two years ago was my twenty first birthday, a Sunday. I went to St James's for the morning service and in the evening took Miss Chattie and Winnie to St George's. The soloist for the anthem 'Seek ye the Lord' was, as usual, Arthur Tinsley. Here I did nothing, not even an extra peg.

Thurs. July 4th

Went to Hannen for the day and found some mistakes in his paragon stove and put them on the right track.

Fri. July 12th

Anniversary of the Battle of the Boyne. Six years ago I had just been two days in Ireland and the Dad, Mater and I watched the procession from the window of the Grand Central Hotel.

Sun. July 14th

First Sunday's work of the year. Now the season really begins. Swinley came to breakfast but no one else to tennis.

Mon. July 15th

Went round work but returned at 10 a.m., feeling sick, violent vomiting, then high fever. Got up again in the evening.

Wed. July 17th

A little better today. Went round some plucking with Dunlop. He has decided to give me a trip on the river.

Thurs. July 18th

Left at 6 a.m. with Godwin. Waited till noon, but still no steamer, then heard that she had run on the rocks at Silghat. Decided to go to Silghat by the down steamer and catch the following day's there. The *Duffla* arrived, no one but myself on board. Reached Silghat 7 p.m.

Fri. July 19th

Woke at 4.25 a.m. in time to get off the steamer and go on the up steamer, the *Sikh*. Mr and Mrs Campbell and Henderson of Nowgong, who were on the wrecked *Mishmi*, were on board. It must have been a very nasty affair. We picked up Edwards at Bishnauth. In the evening anchored between Nigriting and Kokeelamukh.

Sat. July 20th

The wreck seems to have caused general discomfort all up the river. We picked up the Johat padre, Haddo, and a Mr Hainsford. They had been stuck at Kokeelamukh on their way to Dibrugarh and had been obliged to go downstream to Nigriting to get a meal on the steamer. Very cold day for the time of year; foggy all morning, but fine towards evening. Reached Dibrugarh at 6.30 p.m.

Sun. July 21st

Wrote my letters. Colley, the ghat agent, came on board and we had a small talk. Left at 1.30 p.m. and spent the night at Jhansimukh. The evening light on the river was lovely.

Mon. July 22nd

Picked up Henderson at Kokeelamukh and Percy Forbes, one of his assistants, at Nigriting. Reached Behalimukh at 12.30 p.m. and rode Broncho back. Heard that regular night work has started.

Tues. July 23rd

Late wither so we worked till 2 a.m.

Fri. July 26th

Tremendous heat. I had to come in from the garden about 10.30 a.m.

Sat. July 27th

Heat continued and leaf overwithered.

Sun. July 28th

Started rolling at 2.40 a.m. Still no sign of rain. Finished at 6 p.m.

Thurs. Aug. 1st

Went round the new lines plucking and nearly caused a riot by clouting three women, one of whom happened to be Moorali Sirdar's wife. They were plucking into *kapre*, which was strictly forbidden. The punishment was in accordance with Dunlop's ways, but I was too new to inflict it myself. With one accord the women complained to Godwin and Dunlop pitched into me. A few days before he had censured me for not punishing them, so neither way could I do right. Am getting sick of the place and feel inclined to chuck up the job. I can't hit it off with the coolies.

Sat. Aug. 3rd

Piano tuner came over and seemed a capable man. The rains have come to stay: over one and two inches during two successive nights, and today very cloudy. This is the first over two inches.

Sun. Aug. 4th

Very wet morning, but the women turned out to pluck at midday. The piano is finished and Mr Jones gave us some tunes. What a treat it will be to play on now! Last evening

Nando Lal was probing me as to how long I should stay here and the conversation ran thusly:

Will the Sahib return to Belait?
Yes.
Will he stay here five or six years, perhaps?
Perhaps.
As long as the Sahib stays here I will remain. If the Sahib goes and another Sahib comes I shall go the Burra Sahib and get my name cut and go to my country. *Sahib Accha hai.*

I wonder if this flattery is a prelude to a demand for more pay?

Mon. July 27th, 1900 (A year before)

I must now go back and write of the last days in Ireland and at home a year ago. I had visited Mcleod during my holiday and had received instructions about making my application for a billet. So, on my return, I interviewed Mr Pratten and, obtaining a certificate, forwarded it to London, but had received no answer to date. However, some premonition told me it would come today and I looked forward to it all the way back to Bangor. It was short merely saying, 'Call at the office on Tuesday'.

Sat. July 28th, 1900

Applied for leave to go to London and very grudgingly it was given. In the evening walked to Clandeboye and received the good wishes of the Barretts for my success.

Sun. July 29th

Took Miss Chattie to church in the evening.

Mon. July 30th

Did my office work and at 6 p.m. met Miss Chattie at the County Down to receive my night things and she saw me off on the boat.

Tues. July 31st

Left Liverpool Lime Street by the 8 a.m. and reached Euston at 12.15 p.m., going on from there to see the Wallace Collection of pictures, which had only just been left to the nation, at Hertford House. Most noticeable were the Greuzes. There was also a fine collection of armour. At 3 p.m. arrived at McLeods and was offered a billet at Tarajuli Assam if I could leave in a fortnight. I decided to accept and risk any trouble at Harlands. Russel, one of the directors, was with McLeod. He is rather 'oriental' in face and told me he had stayed out there thirteen years without coming home. I hope my fate won't be so hard. Left Euston by the 5.30 p.m. and crossed by Fleetwood.

Wed. Aug. 1st

Got to the office on time and saw Pratten, who was very considerate and told me I could leave at once. Told everyone in the office and my old mates in the shops, receiving general congratulations. Packed all my things and departed like a whirlwind. Lunched with Walmsley and Hayes at our old haunt, the XL Cafe. Returned to Bangor to start packing, having decided to leave Ireland on Friday night.

Thurs. Aug. 2nd

Decided to give a dinner to my Harland and Wolf friends.

My last night in Bangor: how I have come to like the place, and what a pleasant time I have had! Went for a walk round the windmill and have come in and am sitting at the window, but the light has died over the Antrim hills. Whitehead is hidden, the water is dark and there remains only prosaic bed. But a very comfortable place all the same.

Fri. Aug. 3rd

Spent the morning in Belfast. Then, at 6.30 p.m. dinner began with Bell, Walmsley, Hayes, Baker, Hind and Grant. Mickey was invited but did not come. Bell made a speech. I replied and then he had to leave. The others saw me to the boat, where Mickey turned up and we had a parting drink in the smoke room. Mickey making a characteristic speech which astonished some strangers. As the boat left they gave me three cheers. I hope I shall see them again. We drifted down past the island where the arc lights were blazing and out beyond the Twins; then past Holywood and finally Bangor Bay and we could hear the jingle of the hobby horses. The *Caloric*, which had brought me to Ireland, was taking me away. I was 'exceeding sad' to leave a land where I had spent the happiest time of my life, though in truth there were black pages in it. But these only served to make the rest brighter. I sat on deck with the Blayneys and my arm round wee Winnie. How I had worshipped that child, only to find her 'wanting'. But I forgot all against her that night and if I see her again she will be a woman.

Sat. Aug. 4th

Slight fog on the river, which rather spoilt the view. Duncan Peterkin met me at the stage and went off to breakfast with him. Had a farewell talk, 'revolving many memories' of many walks, many drinks with good chaps, many theatres and many flirtations. Left Lime Street at 9.45 p.m., Euston 2 p.m. and got the new Bournemouth express and reached home at 5 p.m. Spent the evening in the tower room. Smoked my new pipe and imagined the lights on the pier were the Bangor hobbies.

Sun. Aug. 12th

Last Sunday at home. Church as usual.

Tues. Aug. 14th

Left for London by the 9 a.m. from the Central. Went to McLeods for tickets, etc. Met Harry and a Stover girl at the Wallace Collection. Dined at the Holborn, then returned to the hotel. Sat and talked for a while in the parents' bedroom and looked out over Trafalgar Square, then turned in.

Sun. Aug. 11th 1901

Rained all day.

Mon. Aug. 12th

Went to Hannen to see his engine, but it is not working well. Hannen drove me back.

Tues. Aug. 13th

Went back again with Hannen and fixed his engine up.

Mon. Aug. 19th

Went through some plucking in the morning but felt sick and giddy. Returned to the bungalow and took it easy.

Sun. Aug. 25th

Dunlop and Godwin went to Gingia for breakfast but returned for dinner, bringing Hannen, Forbes, Hext and Garlant. A considerable amount of bridge was played. Tea house closed about 11.15 p.m. and I made a small distribution of rum, which is drunk even by the kiddies.

Wed. Aug. 28th

Another wet morning but an afternoon wither for part: great quantity of leaf in hand. Worked till 1 a.m.

Thurs. Aug. 29th

Complete wither and started rolling at 4.25 a.m. Midday heat intense and 12 o'clock leaf ready by 7 p.m. so we rolled some. Make nearly 100 maunds of tea of 345 of leaf. I took tea house till 11.30 p.m., then Godwin finished about 5 a.m.

Sat. Aug. 31st

Leaf coming in rapidly. Daily average, including Gingia, about 170 maunds. Tea house closed 10.30 p.m., but leaf spreading continued till 1 a.m.

Mon. Sept. 2nd

Took the tea house entirely by myself and managed it all right. Early finish, 8 p.m. Dunlop and Godwin went to Gingia for dinner.

Sat. Sept. 7th

Woke up to find a regular cold season fog, the first we have had yet, though the morning was not cold. Dunlop went down country.

Sun. Sept. 8th

Anniversary of my arival at Monabari. How well I remember it, and the walk from the ghat, Godwin's horror at my not wearing a vest and my disgust at finding there were no regular factory hours!

Sun. Sept. 15th

Mr and Mrs Macrae came over to breakfast and we opened the new tennis court formally, entrance being made by the Macraes and Dunlop in the buggy. First set Mrs Macrae and Dunlop v. Macrae and Godwin. Court plays well on the whole. I had unlimited labour putting it down and it was levelled all over with a spirit level.

Sun. Sept. 22nd

Godwin went early to Rungaghur for breakfast, intending to sleep at Bargang for shooting on Monday. Dunlop went to Tezpur and I went with him to the ghat and drove Rail back, which was a great concession on his part. Rail is a big buggy horse and christened by the syces, who say he is as fast as a railghari or train. Finished my day in the tea house.

Sat. Sept. 28th

Left at 10 a.m. with Godwin for the Patabghur gymkhana. We drove Sirung and went via the government road, most of which was new to me, and I saw the Borigang tank, which dates back a long way. It was full of *busti* buffalo when we passed. At Borigang we changed horses and were the first to arrive. Tiffin was very good and we all did justice to it, but the gymkhana was no good. First event won by Lillifant, second Godwin, whose jump would really have been the best if he had taken off right. Second event 300 yards race, handicap for age, Malcolm and I scratch, won by Lillifant, second Dr Smith. I managed to finish, which was more than some did. Third event polo ball race, won by Edwards, and the fourth event lime cutting, won by Dr Smith. Then polo was played. Godwin and I went to Borpukhri for dinner and spent the night there.

Sun. Sept. 29th

Mrs Macrae played for me after *chota hazri* and I sang, amongst others, 'Margarita'. At least it was sight reading, as I had never tried it before. Godwin and I left about 10 a.m. and rode to Mijica, where we had a drink with Lecky. Then to Kolapani where we changed to the buggy, reaching home for breakfast. The day closed in the tea house.

Wed. Oct. 2nd

Piari, one of my favourite tea house boys, is dead. I was having a cup of tea after having seen the new bungalow, when Akali came and said the Kirani wanted me, and when I arrived in the tea house I saw a crowd round No. 1 paragon. They made room for me and there lay the poor little chap. He had been cleaning out the machine and some one had shut him into a temperature of over 200°. The skin was burnt off his face, knees and feet where he must have touched the trays in his struggle. What an awful death! I was nearer crying than I have been for years. He was always good at his work and never met me without a smile. His fellow workers looked on absolutely without emotion. I sent a mounted syce for Dunlop who returned about 7 p.m. and sent for the police.

Thurs. Oct. 3rd

The darogha (a policeman of about the rank of sergeant) arrived about 11.30 a.m. and made a lot of useless inquiries about the accident. But every coolie denied having shut the door and we personally don't wish to know, so the matter ended.

Thurs. Oct. 10th

About 300 maunds of leaf stored. Very busy day. Rolled all but the centre of the leaf house bottom floor. Godwin worked from dinner till 1.30 a.m., then I finished at 4.30 a.m.

Sun. Oct. 13th

Usual work. Dunlop says I am to have a spell of garden work this week. I have joined the Assam Valley Light Horse.

34

Thurs. Oct. 17th

Beginning to feel better for my garden work and daily ride.

Sat. Oct. 19th

Home mail. One from Ed, saying he is going to a theological college at Wells for a year, then is going to be ordained. Also a letter from Miss Chattie. The children seem to be all right.

Wed. Oct. 23rd

Dunlop seedy, thinks he has slight dysentery.

Sat. Oct. 26th

Late every night, 12.40 a.m. the earliest this week. Godwin seedy, low fever.

Sun. Oct. 27th

Dunlop drove me to Borpukhi for a service. Edwards, Lillifant and many others present. The usual longwinded sermon from Endell, then breakfast and tennis. Left at 5.30 p.m. and called at the Smiths. Heard the zonophone, which played 'The Holy City' very well; also 'Oh, that we two were maying', and several others.

Mon. Nov. 4th

The Swinleys, Macraes and Garlant came to breakfast and we have some good tennis after. Swinley says it is the best lawn in Assam.

Fri. Nov. 15th

Left about 9.15 a.m. with Dunlop for Patabghur. We rode to Borigang and drove Bheel from there. When we arrived the Tezpur men had not turned up but did so shortly after. They went in and scored 220. Percy Briscoe made 130 before being run out. We made a few runs, then stumps were drawn. Most of us slept in Lawes' bungalow, which is very large, built on a high *chung*, his office being underneath. His bathrooms have piped water.

Sat. Nov. 16th

Chota hazri, then walked to the ground with Allanson, as there were no vacant buggies. Our side were soon out for about 90, of which Lillifant made 58 not out; then a follow on ensued and after a few wickets fell we had lunch. On going out again the side went to pieces with a total of about 70 so Bishnauth lost by an innings and some runs. I scored most of the time and Dunlop umpired in great style, giving sometimes eight balls to an over. At 4 p.m. polo commenced. Bishnauth team 1. Dunlop 2, Hannen 3, Lillifant back Edwards. The match was a draw, Hannen doing most of the work for us, and Eric Hannay for Tezpur. It was a very fast game all through. Dinner was rather quiet but afterwards the furniture moved round cheerfully in a 'Tread-on-the-tail-o'-my-coat' and 'Follow the Man from Cook's' affair. There were many honourable scars but no rows or anything disagreeable. We had a good deal of singing after dinner in which Bishnauth took a prominent part; Tezpur could hardly raise a man. Davidson sang 'Stalls and Gallery', Godwin 'Mush Mush' and myself 'Brown of Colorado'. Crutwell, Bull and Noble also sang.

Wed. Nov. 22nd

Late wither and two days' leaf, so I was up once more till 2 a.m.

Thurs. Nov. 26th

Wet again; in fact it rained all night, so no work in the garden or tea house, as it was too cold. Total rainfall for the day, three inches. Had the stove brought in and sat in front of it most of the day. My left eye has got very painful. Expect I shall have to see Smith. Dunlop went into Tezpur with the recruiting sirdars.

Fri. Nov. 29th

Dunlop drives me to Mijica and then to the doctor, who says I have got a bad dose of opthalmia and must stay in the bungalow. Returned to Mijica and slept in Lecky's bungalow.

Sat. Nov. 30th

Mrs Swinley drove me to the Smiths, who advised me to go home as the Swinleys are afraid of their baby being infected, so they drove me to polo and Dunlop drove me back.

Tues. Dec. 3rd

Rode to Smiths for breakfast. Left eye a little better, but the right inflamed from the sun. Smith ordered me to stay altogether in the bungalow and always wear tinted glasses.

Sat. Dec. 7th

May neither read nor write.

Sat. Dec. 14th

Dunlop moved to the new bungalow. I didn't do much as my feet and legs have knocked up badly. First meal, dinner, at which Godwin appeared in dress clothes, myself in Norfolk jacket and Dunlop in dressing gown.

Sat. Dec. 20th

Godwin went to polo. Packing finished and tea house closed. Rum given out.

Wed. Dec. 25th

Christmas Day, and what a dull one! I stayed at home for fear of infecting someone and my feet were still bad. Hannen and Garlant went up river to ship the ponies for the polo match.

Tues. Dec. 31st

Dunlop and Hannen returned having won their match. New Year's Eve we played bridge to pass the time and were thankful when 12 o'clock arrived.

Sun. Jan. 5th, 1902

Dunlop drove me to Bargang to see Dr Smith and asked

him plainly if I am strong enough to stand next rains, but got no decided answer.

Sat. Jan. 11th

Got all my stores sorted out. The Behali Smiths came to breakfast. Godwin went down country. Dunlop and I paid all the coolies.

Sun. Jan. 12th

Some snake charmers came through the garden and gave a performance at my bungalow. It was rather interesting. Their cobras were fine specimens, about four feet long or more. They also had one of the snakes that the natives say has two heads because its tail is blunt. A small crowd of syces collected to watch.

Mon. Jan. 14th

Took down my line shafting. Swinley came over to meet Mrs Forbes. Having heard a tiger had killed a cow down by the river, Godwin went down and built a *chung* in a tree, but after waiting till 8.30 p.m. saw nothing.

Tues. Jan. 15th

Godwin spent the whole day in the *chung* but though he heard the tiger didn't get a shot, so gave it up as a bad job.

Fri. Jan. 18th

Took another table to pieces. Rode down with Godwin to

the brickfield as I am to take over the job. Uniform arrived so I put on the mess gear after my bath and Dunlop, happening to look in, took me to dinner.

Sat. Jan. 19th

No pony being available I walked to the brickfield and back, this being the most exercise I have had since my eye got bad. The Behali Smiths breakfasted with Dunlop, the latter going to Partabghur for polo.

Sun. Jan. 26th

Straightened up my room in the morning. In the afternoon tennis, at which I can run a little better. After tennis, as we sat on the verandah, a most appalling clamour reached our ears from the garden, a coolie shouting as if in a hopeles state of terror. Godwin, thinking it might be a tiger, sent Sadow, who had been picking up balls, to inquire. He returned and said a man could hear a tiger growling in the *hoolah* on No. 9 but was naturally afraid to go near. Godwin at first decided not to go, knowing that I am not much of a shot, but changed his mind, so we both started in more or less a state of funk, tempered by excitement. We crossed the pruned tea in a straight line from the bungalow and came to the path running from the tea house. On the other side was unpruned tea and the *hoolah* lay about ten yards from the path, being also planted down the sides and bottom with tea. We halted and could distinctly hear growls, but they seemed to come from the top. We peered down the lines of bushes but could see nothing. Then Godwin went a few yards into the bushes and made signs he could see something, but a moment later said it was only an old cow stuck in a drain. And so it was, but her groans were just like growls. We had a good laugh and then tried to extricate her, but failed.

Puncham Chowdikar arrived shortly, having come out after the tiger, with a spear and axe only, and with some others raised her up.

Wed. Jan. 29th

Dunlop drove me to Kolapani polo and on to Mijica for Godwin's farewell dinner. Mrs Forbes, the Macraes, Brown, White and Hannen were there, most of us sleeping in Lecky's bungalow. Swinley made a flattering speech to which poor old Godwin, who was awfully nervous, replied. After dinner we had the usual concert, then musical chairs, which ended in Garlant and I smashing one. Mrs Forbes won the first game, after having sat on Dunlop's knee.

Thurs. Jan. 30th

We intended to get away early, but didn't leave before 10 a.m. It was bitterly cold and we wore overcoats. I went round the pruning with Dunlop and after breakfast started an extra verandah to the stables to break the force of the wind.

Sun. Feb. 2nd

Our farewell *tamasha* to Godwin. At breakfast we had the Macraes, Swinley and Lecky, Chichester, Crutwell, Lillifant, Garlant, Hannen, Dr Smith, Dey and ourselves, in total fourteen. The tennis was not very good, though we indulged in new balls, but the golf course, newly laid out, was a great success. After dinner we had a singsong, Garlant singing 'Cleansing Fires' in his inimitable style. Mrs Macrae worked spendidly at the piano and the evening went off very well. Crutwell slept in my bungalow, my first guest.

41

Mon. Feb. 3rd

Godwin went to Behali and the buggy brought Russel.

Tues. Feb. 4th

Russel did the garden and as usual was not satisfied. Men from Behali brought steel shaft.

Wed. Feb. 5th

Went to the brickfield in the morning. Godwin returned after breakfast. He was a dull white colour and looked quite ill. I have never seen a man so cut up at leaving a place, especially as he is going home. It seems strange. Dunlop drove in Painter.

Thurs. Feb. 6th

Russel and Painter left and Dunlop confided his woes to me. Russel seems to get more extravagant in his demands and this year we can hardly pay any ticca; he has cut down the estimate so much.

Mon. Feb. 10th

Decided about houses in the new lines, relined the shafting and went to the brickfield.

Fri. Feb. 14th

Got my things packed up for camp, taking the *khansamah*, Snoo syce and a grass cutter with me, leaving Nano Lal in

42

charge. My belongings fill a good-sized gharry. Arrived at the ghat at 3 p.m. but no steamer came, neither could I get any food, so had to doss down in the floating dak bungalow, dinnerless and cold.

Sat. Feb. 15th

Woke early and managed to get tea. While I was drinking I heard that the steamer was coming. On her arrival at the flat found myself accosted by Felce. Got my things on board, assisted by the sergeant-major (Danter). Had *chota hazri*, which I wanted badly, having been eighteen hours without food, and gave up keys of Viceroy's shield box to Lloyd. The boat was pretty full and we were quartered on the aft upper deck, where the coolies usually are. After *chota* a little drill in saluting, etc., was given, then we loafed for the rest of the day on our beds, as there was no other room. Heard from Felce that Edridge was on board and was introduced. Liked him very well for first sight. Other chaps I knew were Koch, Percy Forbes, Davidson, Anderson, Duguid, Butcha Hannay, Keith Leigh and Maurice Clarke, who recognised me and came and spoke, though I couldn't remember him till he mentioned Mijica. At every ghat we embarked volunteers and at night time were so full up that there was no room to move between beds.

3

LIGHT HORSE CAMP AT DIBRUGARH

Sun. Feb. 16th

Spent the morning getting into my uniform and fixing things up. We landed about 4 p.m. at Dibrugarh and marched up to the camp. I felt rather strange in my military saddle with sword and carbine but Sirung was all right. We halted and dismounted for a short time opposite where the ghat was in the rains, to allow the syces to come up, and then marched on through a long Assamese *busti*, where there were plenty of pretty girls, and so on into the station. We were preceded into camp by a band and it was all very important. The camp was pitched between the fort and the river. We were shown our tents and told to get into them, not knowing how many men were to go to each. However, Duguid and I, being an odd number, took one and remained in it without additional numbers. it was bitterly cold and as we had nowhere to sit, our gear not having come, we didn't enjoy ourselves particularly. About 8 p.m. we went to mess, which was an ek dum affair, and afterwards put our tents in order (our things having arrived) and went to bed undisturbed.

Mon. Feb. 17th

Fell in for dismounted parade about 7.45 a.m. and did a little carbine exercise, etc. Then *chota hazri* and about

9.30 a.m. mounted parade. I managed fairly well on the whole and at the close, about noon, got some idea of the words of command. At breakfast we were appointed to our respective tables, F Troop being at the far end of the tent. Afterwards I had an hour recruits drill, then my bath, and having picked up an E Troop man, went to the polo ground and watched an inter-troop match. We dressed properly for mess, at which we had the band of a native regiment, and afterwards I turned in early, as there was nothing on.

Tues. Feb. 15th

Early parade and sword exercise. I didn't manage very well, owing to Sirung being frightened and shifting about. The band was present and we had an informal inspection, all over at noon. In the afternoon I went and found young Granville Smith, who is a corporal. I had not seen him for at least six years and the meeting was rather amusing. I was conducted to his tent and found him lying down. I said, 'Are you Corporal Smith?'
    'Yes.'
    'Well, you are not the Smith I am looking for.' (He seemed so altered.)
    'I am the Corporal Smith.'
    'Of Doom Dooma?'
    'Yes.'
    'Then you are the man I want. I am Frank Hetherington.
    After which we shook hands and had a good yarn, only disturbed by recruits drill. However, we met again at the polo ground and he introduced me to several men, also the Rev. Millet, who has a very pretty wife. He comes from Suffolk and knows Rider Haggard. I walked back with a Nowgong man called Allman and after dinner there was a singsong, in which Davidson predominated.

Wed. Feb. 19th

Field day. We fell in about 9.30 a.m. with haversacks and water bottles, etc., and marched direct into the country, our section under Sergeant-Major Koch. Did some dismounted work but didn't see the enemy, so we mounted again and went as far as the range, where we made a proper attack with ball cartridge (the first time on a field day), opening fire about a thousand yards and finishing at two hundred yards, when we charged up and looked at our handiwork; twenty five per cent of hits, which was very good. We loafed about for half an hour while the inspecting officer made a few remarks and then remounted and did some fast skirmishing, where old Sirung came to the front. A little more dismounted work, then we went into camp at 3 p.m. There was a dance in the evening which I watched for a while, then bed about 12.30 a.m.

Thurs. Feb. 20th

Musketry day at the range. First four men of each troop to fire at appearing and disappearing heads; second officers' revolver practice; third, some other target shooting; fourth the Lloyd Lindsay. I watched A Troop go down and then left with Pringle, as we were non-performers.

Fri. Feb. 21st

Inspection Day. We marched to the polo ground and did the usual marching past. Sirung got frightened by the band, bicycle corps, etc. I couldn't hold her. Finally my left spur got entangled in another man's stirrup leather, the ponies pulled apart and my stirrup leather came bodily away. I got mixed up with another troop in trying to get off the ground, but there was no catastrophe. Then Trumpeter Davidson picked up my stirrup, and as I

46

thought the iron was broken upon the saddle, I went off the ground, not sorry on the whole. The rest were not long after me. In the afternoon I studied life in the *busti* and in the evening there were speeches at the club. We went down after dinner to hear them, marching behind Sergeant-Major Anderson's bagpipes.

Sat. Feb. 22nd

Another field day. We marched out to the range and then skirmished up to the railway, where we dismounted and remained idle for a time until a native foot regiment came up. Then we advanced and attacked a *busti*, having taken which we returned home about 1 p.m. In the afternoon I went for a long walk towards the ghat and returned through the Assamese village, talking a good deal with some of the people. After a dinner was a concert, but I went down town in a ticca gharry with Allman and we had a good spree.

Sun.. Feb. 23rd

Arms inspection about 7.30 a.m. Then packed up and at 9.30 a.m. church parade. The Dibrugarh kirk was the first I have been into out here and I was surprised to find stained glass windows. The service was a shortened form and the sermon fairly good. Service over, some went to the upper ghat by rail and on by feeder steamer. The rest, including myself, rode to the ghat, where we arrived and breakfasted. There was not such a crowd as when going up, a lot of chaps having stayed behind. I recognised an old Edmonstone chap named Carless and had several drinks with him. There was also another old Haileyburian, a married man. We reached Kokeelamukh and spent the night there.

Mon. Feb. 24th

Directly after leaving Kokeelamukh we stuck on a sand-bank not 200 yards from the flat and didn't set off till 2 p.m. I lay on my bed all day, the worse for Saturday night's mixture of beer, whisky and cherry brandy. The steamer ran all night and at 10 p.m. I reached Behalimukh and, not feeling like going back, spent the night in the dak bungalow.

Tues. Feb. 25th

Arrived on the garden about 8.30 a.m. and went to the burra bungalow, where I found Dunlop had gone to Rungahgur and would not be back till the evening. Had *chota hazri* and went to the office. At 6.30 p.m. Dunlop returned and I went down for dinner and a yarn.

Wed. Feb. 26th

Went to brick kiln and checked *mistries'* work. Dunlop went to polo.

Thurs. Mar. 6th

Went to brickfield and decided to stop for a day or two to dry the bricks more thoroughly.

Sun. Mar. 9th

As the polo team were to dine here on Monday Dunlop and I settled things and at 11 a.m. he left for Mijica. I went to the brickfield and decided not to shut up on Monday.

Mon. Mar. 10th

Got everything ready in the stables, bungalows, etc. Had breakfast and rode to Kolapani. The Nowgong team were Dr Price, W. Lyall, Percy Forbes and Mackay. Our team were Swinley, Dunlop, Brown and Garlant. Result: Nowgong 3 goals, Mijica 2 goals. Mr and Mrs Macrae, Dr Smith, Lawes, Cowburn, Davidson of Bargang, Noble, Lecky and Butcher were there. The Macraes, Smith, Lawes, Swinley and Brown went back but the rest came for dinner and the night.

Tues. Mar. 11th

Had *chota* by myself and started my *mistries*, the others going to the burra bungalow. Afterwards, knocked round with Lecky. Then he, Cowburn and Lyall came to my bungalow for a drink and we were joined by Dr Price and Dunlop. All breakfasted at the burra bungalow. Then Mackay, Price and Forbes went to the ghat and Lyall with Dunlop to Borpukri.

Wed. Mar. 12th

Dunlop returned to breakfast, saying the Macraes' bungalow was burnt and practically all the furniture destroyed, nothing being insured. They are staying at Dr Smith's till a basha is ready for them. It is very hard on Mrs Macrae, having got the bungalow so nice.

Thurs. Mar. 13th

Went to Kationabari with Dunlop and found the kiln burning. We also went round the hoeing, nurseries, etc. Everything is coming on fast, owing to several little showers.

Paid all the coolies in the evening, which is a good job completed.

Sat. Mar. 15th The Ides

Brick kiln closed. Coming back from Kationabari, Sirung shied in the Kachari lines clean off the road and threw me, but I fell light and wasn't hurt.

Wed. Mar. 19th

No polo. Heard Swinley's coal has arrived at our ghat. First day's plucking, about three and a half maunds.

Thurs. Mar. 20th (*Mimi's birth*)

Started running my engine and found all right. Then went to the coal ghat and had a peg with the captain, returning to breakfast. Dunlop had gone to inspect the local board road at Behali. The captain (Young) walked up in the afternoon and I showed him round the tea house. He told me that he went home some years ago on the *Chusan* and that Harry* was Fourth Officer then, and he had spoken to him several times.

Sun. Mar. 23rd

Rain and very cold. Dunlop had a fire. About 11 a.m. it cleared up and Butcher came to breakfast with rather a tough yarn about two English girls staying with Morris Briscoe at Borjouli. I left at 4 p.m. for Mijica to stay with Lecky and shoot my course at Parbhoi.

* Commander H.R. Hetherington, R.D., R.N.R.

Mon. Mar. 24th

Had *chota*, then drove to Parbhoi and found Mawson and
Mrs Lawton. I think she was Mawson's daughter. Walked
to the range and commenced 100 yards, where I scored
22 points out of a possible 28. I was too sore to fire at 600
and had to postpone it. Lecky was quite satisfied with my
performance. We returned to Mawson's bungalow, where
we came across Lawton, who is a queer specimen. I
breakfasted with Lecky and got back about 3.30 p.m.
Dunlop having gone to Bargang and everything quiet.

Wed. Mar. 26th

Dunlop went to polo. I gave an audience to one of the
Three Graces; they were Rania, Lachaminia and Parbitia,
and used to dance.

Thurs. Mar. 27th

Got an early chit from Dunlop, saying that I have to go
and shoot the remainder of my course. Arrived at Mijica
to find Lecky had not gone. Had some *chota* with him and
went on by myself to the range where I found Fisher,
Lawes and Davidson. Fisher was going through his revolver
practice. I made 15 out of 28 at 600 and 12 out of 14 at
attacking practice. Lecky was off and though he fired 22
shots couldn't hit the target once at 800. There was a drill
in the afternoon at Dr Smith's but I didn't attend, breakfasting at Mijica instead and returning in the evening.

Mon. Mar. 31st

Restarted plucking and get seven and a half maunds.

Wed. Apr. 2nd

Dunlop had a row with the Goungusti, headed by the dak wallah. Having been sent to the office they arrived looking innocent enough but after Dunlop hit one for calling him a liar, out came sticks previously concealed and the four of them wired in. But after a vain attempt they turned tail and ran and men were sent to head them off the factory. Dunlop drove me to polo where Mrs Macrae and Lillifant turned up.

Thurs. Apr. 13th

Dunlop, having wired Cole regarding the Goungusti row, went into Tezpur.

Sat. Apr. 4th

Dunlop returned and said he could make something of his case.

Tues. Apr. 8th

Manufacture. Intended trying my new fine mesh sieve for sorting green leaf, but it wasn't ready.

Sat. Apr. 12th

Dunlop left for Tezpur, as the Goungusti trial was coming on.

Mon. Apr. 14th

Cold and wet. Started building a well at the tea house.

Wed. Apr. 16th

Manufactured part of leaf. Doorgacharan stripped Rago naked in the lines and hammered her. Nando Lal's account of this was very illuminating and took place while he was drying my feet after my bath.

N.L: *Bahut goomal bungalow line me huzoor. Doorgacharam uski maiki mara hai ek dun lingta pingta.*

Self: *Khai ko mara?*

N.L: *Kijani Sahib? Lekin maiki log bahut ghussa hai.*

Self: *Maiki log rhai ko ghussa?*

N.L: *Maiki log bolta ekto maoki deknase sub maiki dekka hai*

A rough transation is as follows:

'There has been great trouble in the lines, your honour. Doorgacharan stripped his wife completely naked and beat her.'

'Why did he beat her?'

'I don't know, but the women are very angry.'

'Why are the women angry?'

'The women say, if you have seen one woman naked you have seen all women.'

Thurs. Apr. 17th

Dunlop returned without being very successful. One man got six weeks, but the rest are to come back.

Sun. Apr. 20th

Some of the women turned out to fill up *haziras*, so I went round them and avoided the farce of a service at Mijica.

Tues. Apr. 22nd

Stopped work at the kiln in order to press on with the

weeding. Feringhu hammered Punia for giving him her opinion of his behaviour with Chakitia, but Dunlop did not take it up.

Sun. Apr. 27th

Spent the morning bottling whisky and completed two casks. No one came over to tennis as it rained heavily.

Thurs. May 1st

First day's sorting with about eleven women. Sorters worked well but needed alterations in one or two places. Well finished with five feet of water.

Fri. May 2nd

Manufacture. Dunlop went to Mijica.

Sun. May 4th

Finished bottling whisky. Hannen came to tennis but no one else.

Mon. May 12th

Went to Behali for golf on the new course and then up to Bargang for tennis. Dunlop stayed the night.

Tues. May 13th

Small manufacture. Paid old lines men and women.

Tues. May 20th

Manufactured a little but greenfly is keeping the garden backward.

Fri. May 23rd

As nothing was doing in the afternoon, Dunlop drove me to Mijica. Only Garlant and Hannen were there and the former sprained his ankle playing tennis. A thunderstorm came on but it cleared up about 8 o'clock and I rode home. Dunlop stayed the night on his way down country.

Sun. May 25th

Hottest day for this year, 93°. Sat in bungalow all day.

Mon. May 26th

Rode to the Gingia Hat to see the well and set up the sinking apparatus. Very hot day so took it easy in the afternoon.

Tues. June 3rd

Found my packer works all right. Rode to Gingia Hat well but owing to a mistake of the *chowdikars* no extra men were there, so I couldn't sink.

Wed. June 4th

Rode to Gingia Hat again and managed to sink a little but as gravel and stones were coming up I saw we had reached the permanent water level. Leaf increasing – seventy-three maunds.

Fri. June 6th

Heard that peace has been declared at last at the end of the second Boer War, but no particulars. The war has dragged on so long that this news causes no excitement here.

Sat. June 7th

Walked with Dunlop to Kationabari. The clearances are full of leaf. He went down country in the afternoon to polo.

Sat. June 8th

Intended to go to Mijica but it drizzled a bit and I didn't feel well so gave it up.

Mon. June 9th

Dunlop went to Soma's *busti* to identify hoes stolen from the garden. Soma and his two wives came in about 6 p.m., accompanied by the darogha. He is absolutely broken down and has given over his *pottah*.

Tues. June 10th

Nanua and Bunsi came in to make their salaams. We have thoroughly reduced that *busti* now that Soma is ruined and the natives may see once more that the sahib can still beat them. Soma is to be allowed to live on his land but we can turn him out any day. Of course the darogha expects and gets a little bakshish for his part in the business.

Wed. June 11th

Dunlop went to Gingia early to arrange about the timber cutting with Swinley. I altered my meals to *chota* at 9 a.m. and breakast at 2 p.m., which I find suits me very well.

Sat. June 14th. The Rains

The rains broke and caused a corresponding amount of *dikh* at the table. Average leaf for the week over 100 maunds per diem so we are ahead of last year.

Sat. June 15th

Wet all day so no leaf was rolled.

Thurs. June 19th

Hannen came over but we didn't play bridge as I had to take the tea house till 11.30 p.m. Gave out grog for the first time this year.

Tues. June 24th

Rather a late star and as everything withers we rolled till nearly 10 p.m. and I finished at 4.30 a.m.

Wed. June 25th

Intended to go to Kettla but was feeling off owing to low fever. Want to make arrangements for boys to go and learn leaf spreading.

Thurs. June 26th

Went to Kettla but fever came on and I stayed the night.

Fri. June 27th

Noble drove me to the river in Dr Smith's old buggy which he had bought; his harness was fastened together with string and the knobs on the shaft, which had broken off, were replaced by nails. However, we arrived safely and I got back by 11.30 a.m.

Mon. June 30th

I rode to Butcher's for breakfast, intending to see the Kacharis work on the local board road. His tea house is *kutcha* but a nice little place and his basha far more comfortable than Hannen's. He does things well too and sported cigars, coffee and liqueurs. After breakfast I got a bad bilious fit and had to lie up till it got cool, when I returned.

Fri. July 4th

Cole came to inspect and found all satisfactory. Dunlop drove him to Mijica for the night on their way down country.

Mon. July 7th

Fixed up governors on the engine and found the packing strips on the crosshead were deficient. Captain Moore arrived in the evening.

Tues. July 8th

Went round Kamjari as Dunlop was entertaining Moore, who had come to inspect saddlery, etc., in company with the sergeant-major of Nowgong.

Thurs. July 10th

First pay day. Dunlop being seedy, I take all old lines.

Fri. July 11th

No plucking. Finished the pay. Trouble in my bungalow which resulted first in Kaddo, then in Nando Lal getting thrashed. The latter broke from me when I hit him about the face with my fists for a while and threw himself over the back verandah railing, falling about eight feet onto his back. He got up, however, and ran off. Dunlop arrived, sent for the chowdikars and found Nando back at his work again. We sent them both to the hospital for the night.

Sat. July 12th

Investigated the case at the office and found them only partially guilty, there having been provocation, so my thrashing was considered sufficient punishment and they were restored to the bungalow once more, instead of going to hoe on hard clay. Left at 10.30 a.m. for Borpukri, sending my marl to Mijica for the night, thinking it would be a good opportunity for the row to cool down, left to itself. Rode Broncho to the Boregang tank and Sirung from there to the Macraes. Their basha is first class and quite different from what I expected. About 5 p.m. left for Mijica and found Swinley in bed with fever and Lecky paying coolies. Dr Smith was there and I salaamed Karoed at the burra bungalow.

Sun. July 13th

Very hot morning so I left Mijica about 10 a.m. When I got back I found Dunlop had gone to Bargang last night. About 4.30 p.m. Dhonesh came up and asked me to go and quieten a row in the lines, as the Mundas had surrounded Pakori's house. Went down and found the Jamadar and setled it. Thakurmoni was at the bottom of it, as usual. In rows both native and European – *cherchez la femme.*

Mon. July 21st

Repaired my paragons.

Tues. July 22nd

We discovered the reason our refiring paragon has worked so badly is that it has drawn air from the front and the temperature has been too low.

Thurs. July 24th

Turning out my old desk I find £2 in sovereigns, which means 30 dibs; this will help to reduce my debts.

Mon. July 28th

Rode to Behali and breakfasted with Smith. He is fit also the missis. Went over to see Crutwell; he has got a very good gramophone. I made him play 'Songs of Araby', etc. When I got back found Dey here and was glad to see him.

Fri. July 31st

Biggest day known on Monabari; 193 maunds and all leaf houses full.

Sun. Aug. 3rd

Sunday plucking, nearly 100 maunds.

Mon. Aug. 11th

Very cold and wet. The women couldn't turn out till 2 p.m. and then only got 50 maunds.

Fri. Aug. 15th

Small disturbance in the leaf house. About ten boys led by Bisram demanded outside work, which they didn't get.

Sun. Aug. 17th

Rained all morning so no wither but at 1 p.m. the sun came out strong and we had to start. All got ready and we rolled till 11 p.m. I closed the tea house at 5.30 a.m., my longest night.

Wed. Aug. 20th

Have now started spreading leaf regularly in the tea house centre *chung* which everybody said was impossible, but I find it gives me a fine early wither if spread thick.

Fri. Aug. 22nd

Dunlop went to Mijica and tried his new pony, Wells, which arrived from Tezpur yesterday. It appears to be very lazy.

Sun. Aug. 24th

Hottest day of the year; 95° in my verandah. Dunlop and I went to Gingia for the day and brought Hannen back and played bridge.

Wed. Aug. 27th

Dunlop tried his new pony at polo but it was too gross to travel fast. However, he expects it to improve. He stayed the night with Hannen

Sat. Aug. 30th

Went round Kamjari and decided to to pluck on Sunday, so made arrangements to send women over to Hannen.

Thurs. Sept. 4th

Mann, the association specialist, and Dr Butler, the government botanist arrived.

Fri. Sept. 5th

Mann went round Kationabari and Dunlop and the others went to Gingia.

Tues. Sept. 9th

Dr Butler returned; also Edwards drove Mann back and as Butcher came over we had a very pleasant evening.

Wed. Sept. 10th

All the party remained so Dunlop didn't go to polo.

Thurs. Sept. 18th

Every possible hand on plucking. Total leaf 195 maunds, which breaks the record. Early finishes every night.

Sat. Sept. 20th

Another big day's leaf. We hoped for over 200 maunds but Kationabari failed and it worked out to 192 maunds.

Mon. Sept. 22nd

Restarted hoeing as the garden needs it badly.

Sun. Sept. 28th

Dunlop returned from Borpukhri and brought Grey. Late night but all rolled.

Mon. Sept. 29th

Rode to Borbheel and, not finding Butcher as I expected, went to Bargang to Chichester and breakfasted. The Beharlites, Noble and Butcher, who had been snipe shooting, arrived about 4.30 p.m. We stayed to dinner, which was a very amusing one.

Tues. Sept. 30th

Left about 9.30 a.m. and met Dunlop and Grey going to Rungaghur. Dunlop told me the packing had blown out of the boiler manhole doors but when I returned they were getting steam up. Heard at Bargang that Godwin had been appointed to the British Assam in Tezpur, gardens Addabari, New Addabari and Balipara, acreage 1200 with two factories. It is a big job for him.

Fri. Oct. 3rd

Made a new arrangement for packing and by 6 p.m. had completed 134 boxes, equal to 157 maunds.

Mon. Oct. 6th

Dunlop and I drove to Gingia for breakfast and returned at 6 p.m. Heard that Lawes has been sacked and that Chichester has resigned. Davidson and Noble have applied for the billet; Noble's opinion of himself is Napoleonic.

Tues. Oct. 7th

Heard that Hannen has been sacked. It must be a great blow to him, poor old chap.

Fri. Oct. 10th

Dunlop brought Hannen over. He looked very much cut up, but it is a good deal his own fault for not taking Dunlop's advice more often.

Fri. Oct. 24th

Dunlop returned early from polo and said the Macraes were coming on Sunday to stay the night; also the Forbes were coming in from the ghat. I had decided to have a trip on the river but in view of the visitors postponed it till Monday.

Sun. Oct. 26th

The Macraes, Swinley, Garlant and Hannen were at breakfast and Butcher arrived about 4 p.m. We had fair tennis and golf. The Macraes and Swinley left about 6 p.m. and the rest of us played bridge until Swinley returned with the Forbes about 10 p.m. One rubber only was played after dinner.

Mon. Oct. 27th

Manufactured some of Saturday's leaf. Left at 5 p.m. for the ghat and on arrival found Mr and Mrs Shuttleworth and a Mr and Mrs Dickson who, I believe, are in the padre line. They took the up boat. Swinley and Garlant turned up and our boat arrived about 7 p.m. with Anderson on board, so we had a good time. At Bishnauth Macrae came on board to see Sanderson and about midnight left again with Swinley and Garlant.

Tues. Oct. 28th

Arrived at Silghat at about 8 a.m. The flat was out in the stream a good distance up from the landing. Fisher arrived in a feeder steamer and we were not detained long. A planter called Hawthorn, with his wife, boarded us but left at Tezpur. I had not previously seen my nearest station in daylight and the view from the sawmills to the ghat was very pretty. All left there except Anderson and a river inspector, Mrs Shuttleworth giving me a very cordial invitation to go and see them when I should next be in Tezpur.

Sat on deck smoking, reading and yarning with Anderson. We breakfasted just after passing Singri Ghat at 2.30 p.m. Reached Rungamati and shortly after Peacock Island came in sight. We arrived at Gauhati just before sunset and the light on the river was perfect, the hills on the far side showing purple and blue and slate through the faint mist. It was the most beautiful sight I have seen since I left home, or rather, my favourite green isle of the sea.

On landing at the flat drove up with the river inspector and Anderson, who wanted some tobacco and, completing our purchases, I returned to the boat, had a hurried drink and departed. I walked to the dak bungalow and found Nando Lal there; also all the rooms engaged by reporters connected with the police commission (which I didn't know was taking place). However, one gave up his room to me, going in with his pal, so I was all right.

At dinner I sat down with two middle-aged men and two about my own age. They didn't know a word of the language as they had just come out from home but had two English-speaking bearers. I altered their ideas of tea considerably and had a very pleasant meal. Went for a stroll with the youngsters but, not seeing much, returned and had a satisfactory night.

Wed. Oct. 29th

After *chota hazri* I went in search of the bazaar, accompanied by Nando Lal. He talked a good deal and informed me that though I had for a short time ceased being his father and mother, now since the trouble in the lines had been settled, I had once more regained that position.

Returning to the bungalow I breakfasted and then got the *chowdikar* to take me round, but didn't see anything worth buying. The steamer arrived on time and I found on board only two men whom I didn't recognise, but saw Glennie in a book so guessed it was the Dibrugarh man of that name, tackled him and introduced myself as Dunlop's assistant. He had been down to Calcutta to meet a young brother about my own age and bring him up to Doom Dooma, where he has got a job. Spent the night betweeen Rungamati and Singri.

Thurs. Oct. 30th

A fog delayed us a little and we didn't reach Tezpur before 10 a.m. One man got on board, Morgan, the river inspector. We arrived at Silghat just after breakfast. It was very different to the first time I was there on a trip. Then I was feeling weak and wretched after repeated doses of fever and the morning was wet and dull with the moist heat of July. I seem to hear the sullen drip of the trees now, but today I was in A1 health with a lovely clear sunny day and cold wind.

It has been growing on me that I have seen young Glennie before. I remember a Haileybury boy in Highfield who had the same curious mouth and boils on his neck, so I asked him and found it correct.

About 6.30 p.m. we reached Behalimukh. Sirung was waiting for me and I arrived back safely to find Hannen and young Main of Main and Co. He seems rather a nice chap. After dinner we played bridge.

67

Sun. Nov. 2nd

Went shooting with Dunlop and Hannen at Kationabari and got four brace.

Tues. Nov. 4th

Rode to Behali to see Anderson about the boiler tubes. He advised to draw them all.

Wed. Nov. 5th

Dunlop went to polo and returned with Perman, Dey and Hannen, but I didn't go to the bungalow, having tea house work.

Fri. Nov. 7th

Dey and Perman came down to the tea house, accompanied by Dunlop who was able to hobble along (he had hurt his back at polo, straining a muscle). Perman advised removing all the boiler tubes. He and Dey left for Mijica.

Sat. Nov. 8th

Manufactured spread leaf so as to leave today's till Monday. Dey and Perman left for the ghat.

Sun. Nov. 9th

Dunlop drove me to service but did not attend himself. Mr Endell preached a thirty-seven minute sermon through most of which I slept. Mrs Forbes, Hannen, Mawson,

Clifford, Glass and Chichester were there. After breakfast Dunlop went to the doctor's, while the rest of us played tennis and the grid. At the latter I established a record. Dunlop drove Clifford back while I rode to Kolapani and Hannen took me on. We had intended a good night at bridge but Dunlop got seedy and left during dinner.

Thurs. Nov. 13th

No tea house work but helped Dunlop with indents, etc. He left about 5 p.m. for the ghat on his way to Bhatkawa Central Dooars, etc., expecting to return in a fortnight. Paid the women and children.

Mon. Nov. 17th

Very little leaf remaining, shall probably not pluck every day.

Thurs. Nov. 20th

All old lines at Kationabari, women plucking, men hoeing, leaf down to 31 maunds.

Fri. Nov. 21st

Stop plucking and put women and children on to thatch cutting. As not enough knives to go round, sent a party of women to hoe at Kationabari.

Sat. Nov. 22nd

Shooting with Hannen at Kationabari. Fair sport; we got

five birds. We left the buggy opposite the bungalow lines and walked in towards my old basha but just as we reached a corner of the tea I spotted something move in front of our feet. I yelled to Hannen, 'Look out!' and his foot had barely touched the ground when he jumped. He must have trodden on the snake's head or somewhere near, as it didn't strike. We killed it afterwards but don't know whether it was poisonous. It was about three foot six inches long, a bright green back and flesh-tinted belly. Hannen left in the afternoon.

Sun. Nov. 23rd

Went shooting in the morning but no result; one bird out of reach, one bad miss and one near home when I had unloaded. In the afternoon I was serenaded by a party of women headed by Romoti and got rid of them with a couple of dibs.

Mon. Nov. 24th

Started all my work and then left for Kolapani. When I reached the polo ground the buggies of the cricket practicers were just leaving and in turning the corner Butcher's pony shied and pulled the buggy over a bundle of thatch, which upset it completely, and Butcher fell out head over heels. Sirung jibbed and I thought was going to turn and bolt but Butcher set his buggy right and shouted to me to come on and I passed him and arriving at the bungalow was introduced to the Guriagan polo team, of whom Hutchinson was skipper. After breakfast Butcher drove me to the field and after Mrs Swinley and Mrs Macrae had arrived the game commenced. There were four chuckers of six minutes and Mijica won by two goals to one.

Tues. Nov. 25th

Got a wire from Dunlop, saying he would return tomorrow, so sent a chit to Swinley, asking him to return Bheel. Flogged Abator at the office.

Wed. Nov. 26th

Sent buggy to meet Dunlop and Macnaught, the manager of Bhatkava. Dunlop was surprised to find the crop not made up but pleased with the valuations.

Thurs. Nov. 27th

Went round Kamjari and Dunlop was satisfied with the work. Told him I had flogged Abator for taking his suckling child from his wife, thus causing her breasts to swell, and for striking her with a hammer. He is not able to work yet. In the afternoon I left for Chichester's farewell dinner at Kettla. Nearly all the district were there, including Glass, and we had a jolly evening. Some of the speeches were very amusing. We shut up about 3 a.m. and I went home with Butcher.

Fri. Nov. 28th

Rose about 9.30 a.m., feeling the worse for wear. Butcher too was pretty bad. Left about 11 a.m. and went up to the bungalow in the evening and found Dunlop very unwell and unhappy.

Sun. Nov. 30th

Finished the indent and packed the balance of an invoice.

Total to date, 5,239 maunds. We may just make the estimate.

Mon. Dec. 8th

Started plucking again with practically all hands.

Tues. Dec. 9th

Leaf satisfactory. Thirty-two maunds.

Thurs. Dec. 11th

Went with Dunlop to Kationabari to start the pruning. I shall like it as it is the most artistic work on the garden.

Tues. Dec. 16th

We are plucking the last time today and expect the estimate is made.

Wed. Dec. 17th

Drove with Dunlop to Garlant's bungalow where a breakfast was given in honour of departing members, after which polo and home. Last manufacture and sorting.

Thurs. Dec. 18th

Packed up everything and realised 5,300 maunds which, being six over the estimate, was very good.

Sat. Dec. 20th

Scarth, the new manager of Gingia, arrived but late, so I didn't see him.

Sun. Dec. 21st

Went to the burra bungalow to get a gun, as I intended walking across the Bargang from Paltan's *busti* to find a suitable place for a road to Borbheel.

After a bit of trouble we reached the Bargang, where we waded across and eventually found a track which took us to Butcher's bungalow. Had a yarn, then breakfast and about 3.30 p.m. left to prospect a road to Monibari, taking two syces. While trying a new crossing Sirung got into a quicksand and sunk up to her withers. I was badly wetted and finished crossing on foot. Butcher went further up to a good place and took the ponies over safely.

Forcing our way through thick jungle, we came across buffalo runs and at one place encountered a herd of them, probably belonging to Nepalis, but half wild. They regarded us in a menacing fashion but did nothiing. We came across a path which appeared to lead to Paltan's *pathar*. Butcher advised me to take it as it was late and he intended going back.

Took the path but it broke up. Kept on for a bit but we did not seem to be going in the right direction so I climbed a tree, but all to no purpose. The sun had set and we were all dead tired, so there was nothing left but to try to find our way back to the upper Bargang. I now began to feel uneasy, as it was getting dark and I dared not risk crossing the river, except at a known place, on account of quicksands. However, I spotted the path Butcher and I had come down and eventually, though it was quite dark, managed to find the proper ford, where we crossed safely and were now on the Borbheel bank. It was impossible to see anything and I began to think we might wander about all night, but as luck would have it we hit on the path leading to the tea and eventually reached the bungalow. I was sopping wet nearly up to my waist, and bitterly cold. So we decided to stay the night.

Thus ended my first experience of the jungle, which nearly had serious consequences.

Mon. Dec. 22nd

Reached Monabari about 11 a.m. and, seeing Dunlop among the pruners, hastened to explain my absence and received fatherly advice concerning the dangers of the jungle. Finished the day in the gardens.

Tues. Dec. 23rd

Went round pruning with Dunlop. Then, about noon, he left for his Christmas trip to Dikorai to go fishing with Lawes and Hannen.

Thurs. Dec. 25th. Christmas Day

Worked in the teahouse and got the engine put together in the afternoon. About 3 p.m. I left for Borbheel via Paltan's *busti*, acccompanied by Nando Lal and Munsi. This time I had no difficulty and rode up the river to where I crossed on Sunday morning, then passed the place where I got lost and finally reached the bungalow without any trouble. No one was there but Butcher and Crutwell soon turned up and we started bridge. Butcher gave us a first class dinner with simpkin, kümmel, port wine, and afterwards we drank brandy and soda with our bridge to ensure a steady head. We had supper and then continued playing until 3 a.m.

1903

Fri. Dec. 26th

We got up about 9 a.m., had *chota* and then started bridge.
The hands were with the dealers and by 11 a.m. I had
cleared my losses and won four rupees which, having
tossed with Crutwell, I raised to eight, and then levanted.
When I got home I found Dunlop, who wasn't up to much,
and neither was I, and so loafed about the rest of the day.

Thurs. Jan. 1st, 1903.

Having made arrangements for the day, I left for Bishnauth polo ground, accompanied by Nando Lal, and
arrived before the majority. The first event was golf driving
and was won by Dunlop. Breakfast was A1 and afterwards
various gymkhana competitions and polo; then we left for
Partabghur, first of all, however, having been photographed and the prizes given by Mrs Swinley, for whom
Edwards called three cheers.

Macrae drove me and I got changed and dressed before
the rest arrived. There was a big party, including Cooper,
who had just arrived at Manori, and Jake Davidson, who
is staying at Dikorai, and the dining room was splendidly
got up. I sat on Clifford's right and Lecky on his left, Glass
being at the head of the table with Macrae and Swinley.
The dinner was excellent but poor Dunlop had got fever
and could not enjoy it. The usual speeches followed, then
singing and shouting and good-tempered horseplay. A few
plates and glasses were smashed and we broke up about
3.30 a.m.

Fri. Jan. 2nd

I got up, feeling all right, and had a big *chota*; then, after
a few words with Clifford, rode home. I then began to feel
the effects of the night before and could eat nothing, so

knowing I looked a wreck I stayed in the bungalow and had tea. At 7 p.m. Dunlop and Hannen returned and we dined in a half-hearted way and went straight to bed.

Sat. Jan. 3rd

I went to Kationabari to mark out a house and see the work at the brick kiln. In the afternoon Dunlop and Hannen went to Runghapur as Chisholm had sent news of a catch of elephants.

Tues. Jan. 6th

Went to brick kiln. In afternoon heard from Dunlop that he wouldn't be home till Thursday morning, as Chisholm had got twenty-seven elephants and the sport was A1.

Fri. Jan. 9th

Rode to Behali to arrange with Dey about the boiler, which he is to superintend and decided that he is to make the gear. On returning home heard that Godwin is expected on Sunday.

Sun. Jan. 11th

Butcher was the first to arrive, then Davidson and Scarth and finally Godwin, driven by Lecky. He was much the same, only looked rather unwell and older. After breakfast we played golf and tennis.

Fri. Jan. 16th

Dey came over to do the boiler and brought his new assistant with him, a non-engineer. We succeeded, after alterations to the gear, in drawing four tubes.

Sat. Jan. 17th

Worked all day on the boiler and drew seven more tubes.

Sun. Jan. 18th

Noble came over from Kettla on foot, as he had been shooting and Hawthorn from Behali. The latter is an unmentionable bounder and immediately on arrival suggested drinks, as though the place belonged to him. He professes to be a rider but his pony's harness was all wrong and its mouth cut and bleeding. After breakfast, tennis and golf, Hawthorn left with Noble after dinner, expressing his dislike of the trouble of returning home, but no one pressed him to stay or ever show his face again at Monabari.

Fri. Jan. 23rd

Gora returned from Tezpur, having been flogged by Brown and kept in futtock on the garden for over a week.

Sun. Jan. 25th

After dinner had a final look at Expectation. He appeared very weak. Gave orders for the syces to wake me if he gets worse.

Mon. Jan. 26th

Was wakened about 1 a.m. and heard Expectation was worse. Went down and found him outside his stall and lying down, too feeble to raise himself. Sent for help and finally got him back on to his bedding. Got up again at 6 a.m. and heard he was dead, which will be a nasty shock to old Hannen. Went down after *chota* and had his tail removed as a memento.

Thurs. Jan. 29th

Dey came over and examined the boiler tubes. They are all badly pitted and cannot be used again.

Sun. Feb. 1st

Dunlop and I went round the edge of the garden from Beharies lines to the seed garden and looked at the land being cleared for dhan. This was the beginning of the scheme to induce the coolies to stay by giving them rice land. After breakfast Garlant and Crutwell arrived, informing us of a tennis tournament in which one garden plays another, so we played them and won, two sets to love.

Thurs. Feb. 5th

Went in the afternoon with Dunlop and Hannen on their way to Borbheel. We followed my track down to the river past Daria's *busti* and I left them opposite Butcher's washing ghat, where they managed to cross. Returned myself with a *busti* wallah who undertook to show me a short cut across the *bheel*, which we thought impassable. We succeeded and it will be very simple to cut a road for the cold weather.

Sun. Feb. 8th

Macrae, the Swinleys, etc., came to breakfast and after-
wards the tournament went on. Dunlop and I beat Macrae
and Scarth two sets to one; then, as I rather expected, the
Swinleys beat us two sets to love, for I played very badly.
In the evening we had bridge as Butcher, Scarth and
Macrae stayed the night.

Mon. Feb. 9th

Macrae, Butcher and Scarth went round the garden with
Dunlop and very much approved of it. After seeing
Kationabari Butcher didn't talk much about Borbheel.

Sat. Feb. 14th

Hannen's last night. Scarth came over and we had bridge.

Fri. Feb. 20th

Russel turned up but I had no conversation with him.

Sat. Feb. 21st

Heard that Russel is as sweet as honey this year and says
the garden is looking splendid. Went up to the bungalow
in the evening and had my peg and a yarn.

Tues. Feb. 24th

Dunlop left for a manager's meeting at Dr Smith's

bungalow, so I entertained Russel and saw him off the ghat. All pruning finished at Monabari.

Sat. Feb. 28th

Dunlop went down to Borpukhri for golf. Kationabari pruning finished.

Fri. Mar. 6th

Sergeant-Major Danter came over on his way from Dibrugarh as Burton has returned and he goes back to Tezpur. He said camp was a great success.

Sat. Mar. 7th

Danter put me through sword exercise and carbine and then left for Partabghur with Dunlop, who was going to polo; he will also make arrangements for shooting.

Mon. Mar. 9th

Rode to Dr Smith's for the night with a view to shooting my course at Pabhoi.

Tues. Mar. 10th

Danter drove me to the range where we found Barker, Duguid and Fulford, who is repairing Macrae's machinery. He failed after 500 yards, Duguid after 600 yards. Only Barker and I finished.

Wed. Mar. 11th

Smith came over to see Satis Babu's wife, who is ill. The old order verily changeth.

Mon. Mar. 16th

Loafed about all day and am feeling very seedy. Think it is want of a change.

Wed. Mar. 18th

Went to polo with Dunlop and as Smith was there, told him I was feeling off. He said low fever and prescribed quinine and Fellow's Syrup combined with a change of air.

Sun. Mar. 22nd

Rode over to Behali against a strong wind to get some advice about the tubes. Arranged that Smith should come over on Tuesday. When I arrived home I met Dunlop opposite the tea house. He had word that an elephant or *hathi* was in the tea at Kationbari and was going down so I went with him, but it was too dark to do anything and we returned.

Tues. Mar. 24th

Smith arrived with his missis and kid about 11 a.m. After breakfast we started on the tubes and soon had forty in, the rest having not arrived.

Wed. Mar. 25th

Started rivetting up and expanding. The Smith party left about 4 p.m. Dunlop went to polo and returned with the Macraes, Miss Adamson, Dr Crow, Edwards and Clifford. We had a jolly dinner and afterwards bridge and stayed the night.

Thurs. Mar. 26th

On reaching the tea house I found my balance of tubes had arrived so started men to file them. Edwards and Macrae looked over the tea house and approved of it. After *chota* the doctor examined me and pronounced jaundice, putting me on a diet, etc. I spent the rest of the day at the boiler. At 5 p.m. Dunlop took me up to the bungalow where tennis was going on. He left about 6 p.m. with Miss Adamson and the rest shortly afterwards.

Fri. Mar. 27th

Smith came over and started expanding the tubes but some gave a lot of trouble and three had to be changed.

Sat. Mar. 28th

Smith came over again and we got on better, practically finishing the smoke box end; one more tube had to be replaced.

Sun. Mar. 29th

Wrote letters and in the afternoon Harnack's man called. I didn't buy anything but told him to hurry up my boots from Morrison Cottle and Co. He spent the night here.

Mon. Mar. 30th

Expanded tubes, two of which had to be replaced, owing to bad flaws. Filled the boiler up and at 5 p.m. another tube burst, requiring the water to be run out. It is getting monotonous and soon there will be no spare tubes left. The welding has been disgraceful.

Tues. Mar. 31st

Smith came over and we started to raise steam, geting to 120 lbs. without a leak. Dunlop returned and we had a drink while steam was raised, then went down and ran her for about ten minutes at slow speed.

Wed. Apr. 1st

Smith arrived and we ran the engine successfully, putting all the machines on. The governors controlled fairly well.

Thurs. Apr. 2nd

I ran the machinery all day at a slow rate to let the bearings bed themselves. We plucked thirteen maunds of leaf.

Fri. Apr. 3rd

Started manufacture with the machinery and all went well. Have arranged to go to Tezpur on Sunday.

Sun. Apr. 5th

Dunlop left for Runghapur and about 4.30 p.m. I rode Sirung to the ghat. I met a Nepali who passed me without dismounting. If ever I see him again I will give him something to remind him. After Behali village passed the Mijica dhobi who is going to his country for a holiday, and at the ghat found our Kirani Babu, on a like intent. I had been there a short time when another sahib turned up and I found he was our new D.C., by name Lees, a rather dull-looking specimen. The boat was up to time and there were several men on board, of whom I only knew the Nowgong sergeant instructor. As I couldn't get a cabin to myself I went in with him. We anchored at Silghat for the night.

Mon. Apr. 6th

Reached Tezpur just before 7 a.m. and went immediately to the tramway station. There was no one I knew on the platform but three men were discussing the cholera outbreak. The train left about 7.30 a.m. and two men got in with me. It was a strange feeling being on a railway again, almost homelike. Between Tezpur and the first garden, Bundukuri, the country was principally *dhan khets* and everywhere a large number of bamboos. At Bindukuri one man got out and I learnt from the other he was the great George Moore, out on his annual winter visit. The other proved to be Bruce, the manager of Nahorani, and he pointed out to me the various gardens and objects of interest, such as the celebrated Tezpur bank of red *matti* beginning at Thakurbari and running out to Addabari, the discovery of which made several men independent for life.

At Phoolbari station I found Godwin's buggy and a note telling me the pony, Five of Clubs, was groggy, so I must look out. Bearing this in mind, I let the syce lead him down the hill from the station (the first I have seen in this country, as Bishnauth is so flat) and soon came in sight of the factory. I drove up the 'bank' again, past the factory and office, arriving in front of one of the best bungalows I have seen.

Godwin appeared and has just the same old energetic manner. He was looking much better than at Monabari last January. After *chota*, walked round the tea house, which is well found in every way. I thought his labour force much cleaner than ours and the women better looking. The garden is full of deep hoolahs and as different from Monabari as chalk from cheese.

In the afternoon Godwin drove me to polo and introduced me, among others, to Moss, The Empire's accountant, Roffey, assistant Bojuli, and Dr Bentley. The Thakurbari ground is very pretty but not as good as our own at Kolapani. Everyone was very hospitable and though nobody stayed late I was three parts full and had to hang on to the buggy, being giddy. When we got back nature asserted herself and got rid of all I had swallowed. It reminded me quite of Belfast. But I dined and slept well.

Tues. Apr. 7th

Went round the garden with Godwin and walked over to New Addabari via the garden tramline. Met Campbell, his assistant, had a drink in his bungalow and walked back, breakfasting at 1.30 p.m. Godwin paid some coolies in the afternoon. In the evening we drove over to the Watsons at Nahorani for dinner (he is super of the BITC) and while passing through the forest found our way blocked by a fallen tree. We managed to get past, however, by taking the pony out. Played bridge after dinner. After four rubbers there was only a difference of four annas between us so we didn't pay. It was a lovely moonlit night so we had a fine drive back, returning a different way.

Wed. Apr. 8th

Went by buggy to the new garden where Godwin was much upset by Campbell's draining work. But he is new; I think Godwin shouldn't have left so much to him. Returned and drove over to Balipara, passing the polo ground, now disused. Then passed through Bamjamy and crossed the Masiri river, which has a pebbly bed; quite a change from our everlasting sand. We found Curtis at home and in due course had breakfast (not very well cooked) and left about 4 p.m. Campbell came to dinner and we played bridge afterwards.

Thurs. Apr. 9th

Walked to the new garden, where Godwin spent a long time teaching the women to pluck and I in taking stock of them. We returned by trolley. I sent off my marl to Tarajuli, having got word from Felce that he would meet me at polo. After breakfast we left, and I felt much regret at going; in fact wishing I had decided to spend all my time at Addabari, but maybe I shall get leave later on.

When crossing the bridge close to the polo ground Five of Clubs shied at some cooli clothes on the rail and we were very nearly into the nullah, but luck was with us. After a little while Felce arrived and as soon as the polo was over we left and had a good drive back, as his pony is exceptionally fast.

We passed Borjuli tea house (the biggest in Assam) with its electric light, etc. Arrived at Tarajuli and had dinner, an exceptionally good one, with Allanson and Edridge, who was very full of what he had seen and done at Studley Tower my parents home. We were not late turning in and my room was very comfortable. This is a *chung* bungalow and Felce has fitted up electric bells all over, which is very convenient.

Fri. Apr. 10th

After *chota* took a stroll into the garden accompanied by Cowburn, who had just arrived, and went as far as the boundary of Dhenri, where the assistant, Reid, came to speak to us. We also saw Townsend, the new forest assistant, just out from Coopers Hill; he was on his way to inspect rubber tapping. Towards evening we went over to Edridge's for tennis. There we met Keats, the new European barrister for Tezpur, and his wife, who play rather well.

After dinner the *chowdikar* brought up a woman who had been caught trying to commit suicide, but Felce managed to make her look a little more cheerfully on life. He has two little Munda girls working in his compound, which whom I fell in love. We pelted them with plantains from the chung and eventually they came up and picked some for themselves. They were not shy or insolent like most of our Bengalis.

Sat. Apr. 11th

Went round the garden, which looks well, but badly eaten by ants in parts; also into the tea house, which is not up to much. Then to see a bridge on the Rangapara road. Sent off my marl to Chardwar and left about 4 p.m. Cowburn drove me to the Borjuli turning, where I got Felce's buggy and went on the road through the forest, which was very much cut up. We crossed one river, the Gabri, and eventually reached Dey's. He was out fishing. He turned up at dusk. After dinner we played billiards on a small table whose cloth was in a terrible state. His gramophone was the biggest I have seen and cost 200 rupees.

Sun. Apr. 12th

Koch and Pulford arrived about 10 a.m. to get a dak on to C.C. Clarke at Deckajuli. The former told me he had just caught a batch of five Arcotti coolies belonging, I think to Phulbari. They consisted of two dhobis, a napet, a brahman and another of low caste. Recruiting is not what it used to be. After they had gone we strolled down to his tea house, a miserable show with a fine engine and nothing for it to drive. Then had breakfast and left about 3.30 p.m. and drove to Bindukuri station, where I caught the 5 p.m. for Tezpur. Went to the dak bungalow. There were only two people there, a visiting dentist and a patient from Nowgong. Almost directly after dinner went to bed.

Mon. Apr. 13th

Rose about 5.30 a.m. and went down to the steamer, which had arrived overnight. Did not leave till 7 a.m., by which time the down steamer came in with Percy Forbes on board to visit the dentist. Scarth boarded the upward. He had been staying with the padre, Endell, and told me he would drive me from the ghat. We had *chota*, then smoked and read in a cabin as the wind was too strong outside. Arrived on time at Behalimukh. We found Dunlop asleep in the bungalow but he soon got up and Scarth having left, we had a yarn. I also dined with him. So ended the most enjoyable trip I have had in Assam.

Wed. Apr. 15th

Rain wanted badly; practically nothing to pluck. Cholera bad at Behali; about sixteen Assamese have died. Dunlop drove to Gingia to inspect sawing and I rode to polo.

Sat. Apr. 18th

About ten people down with choleric diarrhoea. Hope it will not result in an epidemic. Dunlop went down country.

Mon. Apr. 20th

A few more cases of diarrhoea, but only one man dead, Dunlop's mali, Jotia.

Thurs. Apr. 23rd

Manufacture and line building.

Sun. Apr. 26th

Dunlop drove me to Behali for golf. It was the first time I had seen the links and thought them very pretty. We went back to Bargang for breakfast and afterwards had some indifferent tennis. We drove home at night.

Sat. May 2nd

Usual work in the morning. In the afernoon sat in a *chung* in the hopes of shooting a leopard, but it did not come.

Sun. May 3rd

Helped Dunlop straighten out the office.

Mon. May 4th

Dunlop went to polo. The clubs having been amalgamated, there will be alternate Mondays only at the grounds.

Thurs. May 7th

Paid new lines and then played tennis with visitors.

Sat. May 9th

Dunlop went to Behali for golf and on to Bargang for the night. I finished pay and then rode over to stay with Swinhoe to be ready for the game drive in the morning.

Sun. May 10th

Noble roused us about 6.30 a.m. and shortly after we had *chota* and went over to Davidson's, where we found Butcher, Dunlop and Wolverstone, who had arrived at night. When the coolies, about 250, had been mustered we started and took up positions along the Bargang about 100 yards apart. I borrowed a buckshot gun from Swinhoe, a ·303 rifle, also a 12 bore shot and ball from Dunlop, and so was well off. The drive was soon over and resulted in nothing, all the game having broken back. We then went to the road between Bargang and Kettla, sending the coolies round by the ghat road. While picking my way to my position my knee suddenly gave way, as it has not done since the operation, and I couldn't move for some minutes. However, it wasn't very bad and I reached my station. This drive, like the last, proved futile. We all breakfasted with Davidson, then played bridge and finally had a pleasant drive back in the morning.

Mon. May 11th

Dunlop went to polo and returned with Lawes, Brown, Davidson and Butcher.

Tues. May 12th

Met the visitors at the tea house and then did old line plucking. After breakfast Kaloo brought me Dunlop's salaams and I found him at the office with Macfarlane, whom I had met once before. We went up to the bungalow for a drink and Dunlop told me he was going to Bargang for the night with Macfarlane and then on to Nowgong for a few days.

Fri. May 15th

Sent the buggy for Dunlop and was rather surprised when he turned up. Talked things over in general and he seemed satisfied.

Sun. May 17th

Dunlop drove me to service at Borpukhri. There was a fair muster but we had to sing without music as Mrs Macrae's book with tunes was burnt.

Sun. May 24th

I went to the burra bungalow for a drink and while there a *chowdikar* came to report that Aguda had been bitten by a cobra at noon and had died four hours later, which was quick work. The snake was one kept by Behari and the two men had been playing with it. Behari was brought up and acknowledged it, also confessing that he had afterwards turned the beast loose. Dunlop gave orders for the police to be informed and a search party for the snake to be organised.

Mon. May 25th

After *chota* Dunlop and I went to the lines, where we found the snake had been caught by Behari. His wife had been plucking *benjon* outside her house and it had struck at her, so she called her husband who secured it. It was about two feet six inches long. In the the afternoon the darogha turned up and we decided not to prosecute but everyone was warned and Behari killed the cobra in front of us. I went to look at Aguda, but there were no marks to speak of; only the fingers slightly swollen where the snake struck him. This is my first acquaintance with death from a snake bite.

Tues. May 26th

Usual work. The babu has now taken over the tea house. We started packing our first invoice.

Thurs. May 28th

Went round the hoeing and found a good deal of lattera work. My wrath, which had been accumulating, burst forth and I hammered Behari VI. I told Dunlop about this in the evening and presently the man came up to complain. Dunlop, of course, would not listen to him but sent him to the hospital and told me to go and see after it. I went down and talked to the man, telling him he had been a fool and got what he deserved, but now I would look after him.

Fri. May 29th

Sent for Behari and gave him a bottle of rum. It is the best way to deal with a coolie: half kill him when he plays the fool and then show him how kind you can be if you like.

Sat. May 30th

Dunlop went down country for golf. There was no plucking and everyone on the hoe.

Mon. June 1st

Dunlop was seedy and didn't go to polo. Hawthorn came over and breakfasted with us. He also stayed to dinner with Dunlop.

Thurs. June 4th

Home mail and a very funny letter from Duncan about Miss Mendinoca. We could teach him something out here.

Fri. June 5th

Dunlop left for Borpukhri to meet the Prices who are coming for the Nowgong match. They will stay with the Macraes till Sunday and then come to Monabari.

Sat. June 6th

Heard from Dunlop that I had been invited to the dinner and so I sent off my _duds_, as I had no pony (Sirung being at Addabari). I borrowed the kyah's tat and after much flogging managed to make it move. I expected to be the last but, barring Butcher, proved the first and it was nearly an hour before the teams turned up. Nowgong were represented by 1 Footman, 2 Price, 3 Davies, Back Wylie; Bishnauth by 1 Garlant, 2 Crutwell, 3 Butcher, Back Edwards. Dunlop was too seedy to play but Butcher took Broncho to help himself. Nowgong won by one goal.

Noble drove me to Crutwell's, where we changed. We had a good dinner but Noble had to retire early, being full of bass, and wasn't seen any more. After dinner we had the usual sing-song and a little wrestling, but were not late in turning in.

Sun. June 7th

Went over to Garlant's for *chota* and were in time to say goodbye to the team. Then Dunlop drove me back, Price and Mrs Price following. Mrs Price was seedy and on reaching the bungalow retired to bed with fever. I breakfasted there and played golf with Price as Dunlop had hurt himself wrestling.

Tues. June 9th. The Rains

The rains broke about 11 a.m., with a very heavy storm which lasted about an hour, in which an inch fell; then it drizzled till 4.30 p.m.

Thurs. June 11th

The Prices left as Mrs Price was feeling better. Their visit had been very unfortunate on account of Mrs Price and Dunlop being seedy.

Fri. June 12th

Dunlop sent for me in the afternoon and showed me a wire from the agents, saying, 'Clarke dead can you spare H for a few weeks to take charge at Deckajuli.' It gave me a surprise and in a way a pleasant one, as it is a fairly big job for an assistant with only two and a half years' experience and if I don't make a mess of it, my name will stand rather above other assistants in our agency who have not been on their own hook. The garden is about 400 acres and belongs to Perman.

Mon. June 15th

Wire arrived, ordering me to leave at once. This suits my domestic affairs.

4

IN CHARGE AT DECKAJULI

Tues. June 16th

Sent a wire to the Deckajuli babu about daks and I left for
Kettla so as to have no trouble in the morning crossing
the river. I made the tea house women think I was going
for good and Romoti announced her intention of coming
with me. It is rather amusing giving people a surprise, as
you get them off their guard. Noble was in great form and
I had a pleasant evening.

Wed. June 17th

Drove to the ghat by the Nanyhur road, which was a treat,
and was in plenty of time, only to discover I had left my
bag at Kettla. I sent a chit, however, by the syce, telling
Noble to forward it. The steamer was on time and at
Bishnauth I saw Davidson and Wolverstone. They were
on their way up river. Davidson advised me to land at
Tezpur as Dey had taken the ponies, etc., over to Chardwar
and was unofficially in charge. I was in a bit of a dilemma
as I had wired the babu, but luckily Dey met me at Tezpur
and I got off and we went to the chummery for a drink.
Dey's buggy was waiting and we reached Chardwar about
4.30 p.m. Had some tea and took a stroll. I learnt from
Dey that after the coolies had hammered Clarke in April
he at first got fairly well, but had latterly developed

erisipelas of the head and died at Thakurbari. The coolies seem to be a sullen crowd and he was too zubberdust for a new manager. After dinner we played the gramophone and soon turned in.

Thurs. June 18th

We spent the morning going round Chardwar. It took some time as Dey stopped to chuck every woman under the chin. After breakfast we drove to Deckajuli which is about twelve miles. We had to cross the Belseri on a *mah* but otherwise there was no trouble. We went down to the office and interviewed the babu, also reproving him for taking a pony out of Dey's stables. His letter of explanation was as follows:

Deckajuli T.E.
18.6.03

To J. Dey, Esq.
Chardwar T.E.

Sir

I have received your kind note. I will have every attention to carry out your orders perfectly.

About taking the pony from your stable without your order you should kindly excuse me for the reasons stated below.

I sent two wires to you (one to Bindukri and one to Tezpur) but I was not quite sure whether you got them as I had no reply from you. Moreover, I was puzzled about what to do as I thought if you would make no arrangement (I was not sure whether you knew when Mr Hetherington is coming) then what would be the fate of the gentleman (Mr Hetherington) at the jungles of Singri Ghat where there is not a human being to be found. You may easily understand how angry he would have been with me if it was so as above, in spite of his wiring me on Tuesday. It was the mistake of the messenger I sent to ask the syce to bring the red pony instead of the Bhutia pony as I have told him. So you should kindly excuse me for the fault I have done to you and Mr Hetherington.

Yours faithfully
D. Saikia

Fri. June 19th

Dey gave me the keys and then left about 10.30 a.m. so I was for the first time the 'boss.' I went to the office and gave orders about accounts, then inspected the hospital, which is only a *dewai khana*, as the coolies don't expect to stay there, and went into the garden where I picked up the jemader and new Mohurier. The plucking is very hard and I didn't like the appearance of the bushes which are pruned right down the sides. In the afternoon I straightened up the office in the bungalow. I received a letter from McLeods, saying I shall be here a month, then Colin Whyte will relieve me and my pay is to be 350 rupees per month.

Sat. June 20th

Did a job or two in the office and looked at the tea house. There is nothing particular, except Perman's press and an apparatus for breaking up caked lead. He fires at very low temperature and I think kutcha, but I shall not interfere. After *chota* went round the hoeing, which is fair, and also the plucking. The people are sullen and will hardly answer but are ready to carry out orders.

Sun. June 21st

To fill up time drove as far as the Belisiri and back. The buggy is much too heavy for the Bhutia pony and I could have almost walked as fast.

Mon. June 22nd

Went round the work during the morning and found the Kacharies hoeing very bad, if I were to be here longer I would turn them out.

Tues. June 23rd

Packed twenty two boxes of fannings. I have now inspected the various roads round here but as the pony is not strong enough for the buggy there is not much pleasure in driving. Also, as I neglected to bring my music there is nothing to do after 6 p.m., so I shall be deuced glad to return to Monabari. Nando Lal is absolutely homesick; he wants to see his grannie, sisters and livestock, but doesn't care a straw about his wife. It is perfectly true, as he says, that if one dies there are plenty more to be got, but he has an exceptionally pretty one, and like a native, can't appreciate it. Their ideas of beauty are very crude.

Wed. June 24th

A fearfully hot day, My thermometer showing 96°, and this bungalow is full of mosquitoes and little gnats which don't sting but make an infernal row. I have developed a heat rash from sweating and the irritation round my waist where my belt chafes is maddening. One of the brutes engaged in Clark's row came in on Tuesday evening; he was the instigator, though, as he didn't assault, only got two months. I presented him with a name cut chit and cleared him off jat-pat. I have never seen a more hangdog scoundrel, but his wife was a very pretty girl.

Second diary missing

5

BACK AT MONABARI

(Third diary)

Wilt thou take all, Galilean?
But this thou shall not take!
The Laurel, the palm and the pean,
The breasts of the nymph in the brake
And all the wings of the loves
And all the joys before death –
Thou hast conquered, O pale Galilean,
The world has grown gray from thy breath.

Sun. Feb. 11th, 1906

Here beginnith the first page of the third book of the diary of Frank Arnold Hetherington, Gentleman. How like this sounds to the commencement of reading the lessons in church and what a pity that the old title 'Gentleman' is out of use, except in a few cases, in army gazettes, etc! The story of John Halifax Gentleman was in my mind as I wrote this.

There was a shower of rain in the morning so I delayed going to Bargang till 11 a.m. To my surprise I heard that a lot of men were shooting at the butts and that the sergeant-major was there. They arrived after about two hours, Johnny, David and Filkin having completed their course and Robinson still to fire at 800 yards. Davidson got leave for me to stay the night and all the others left at 5 p.m. I was feeling seedy as three days of toothache were resolving themselves into a gum boil. I slept in the same room as the sergeant-major and had a bad night, partly owing to his snores.

Mon. Feb. 12th

Danter and I made an early start and once I had fired a few rounds he saw my sighting was good but I was jerking the trigger – a fact which Davidson had not noted. I endeavoured to remedy this and got on well, my total eventually being 172. Robinson also completed his 800 yards then he went to his bungalow and got our buggy. On the way back saw the doctor and showed him my mouth which was very painful, and he promised to come over on Wednesday. He drove me to Gingia Hat. I drove Mona back and after seeing Dunlop at the office went to bed.

Tues. Feb. 13th

Got up, feeling much better and able to do a good day's work. Dunlop went to Kationabari and, finding a loss made in the pruning by Harirman, decided to fine him.

Wed. Feb. 14th

Dr Smith came over but as my mouth was practically all right he didn't have to operate. Dunlop and Johnny went to polo.

Fri. Feb. 16th

Weather very cloudy and damp. An escaped *hathi* came

102

strolling along during the afternoon but none of our coolies knew how to catch it and it disappeared towards the Bargang. It was quite tame.

Sat. Feb. 17th

Monabari pruning is now nearly finished and some women will go to Kationabari on Monday. Johnny departed for camp and Dunlop to Partabghur for polo. Braddan returned Sirung and the buggy, which he has had for a year.

Sun. Feb. 25th

Garlant and Crutwell turned up about 11 a.m. and we started tennis but the quality was poor. The Swinleys and Ronald Davidsons came to breakfast and taught us a game called Russian Pool, a sort of combination of ordinary pool and snooker but superior to either. The Swinleys and Davidsons left at 6 p.m., while the others dined and played bridge.

Mon. Feb. 26th

Started fixing up domestic affairs and visited Kationabari pruning. At Monabari the women commenced hoeing. After breakfast we distributed *pathar* at the new place near the east of Kationabari and went on to Borbheel for the night. Levick was there and we played bridge.

Tues. Feb. 27th

Dunlop went straight to Behali in the buggy while I biked to Rangsali to inspect some property and then went on to Behali for breakfast. We left about 3 p.m. in order to have a good evening's work in the office.

6

GOING ON HOME LEAVE

Wed. Feb. 28th

Spent the morning packing and got everything completed by breakfast. Dunlop came in about that time and said Buddhoo's hoeing was very bad, which I can quite understand – that man will give trouble yet. During the afternoon we went to polo and Burton also biked over. The usual crowd were there and the customary farewell drinks were stood and mopped up. We dined at Gingia and played bridge after.

Thurs. Mar. 1st

Got my marl dispatched by 10 a.m. and settled accounts in the office, Dunlop giving me a cheque for Rs. 1,100 and cash Rs. 200. I also left Rs. 100 in his charge. I have saved about Rs. 1,800, which is not bad. Had breakfast with Dunlop and as I finished Burton came up and at 1.15 p.m. Bheel and the buggy were brought round and I felt in great spirits. Hirasi pumped me as to whether I was coming back and said people were sorry I was going.

Arrived at the ghat to hear the steamer was late so I made Radhoo unload my marl and let him go. He was quite tearful and in saying how good I had been to them, laid great stress on the amount of leave I had given them. This shows what line to take when I come back.

At 6.30 p.m. went to the floating dak bungalow to make arrangements for the night. However, after half an hour the downward boat arrived and on boarding her she proved to be a big steamer being used for the nonce as the mail. I seemed to remember the captain and on asking if he didn't once bring coal to Kationabari he mentioned my name and said he recollected the visit. There were only two passengers, a German Lutheran missionary and his wife. The two mates dined with us in the saloon. In these boats the captain runs the messing and it wasn't up to much.

Fri. Mar. 2nd

Arrived at Tezpur at 9 a.m., and went to the chummery where I met Cragie, our new solicitor, transacted some business and then took my carbine, etc. to the sergeant-major's. He was not at home but Mrs Danter was and I stayed for half an hour and talked, then returned via the Assam Valley shop, where I paid a bill to the chummery. A wire from Godwin was waiting, saying come out by the trolley. But it was too late and I sent word I would go by the 2.30 p.m. train. At Thakurbari Miss Godwin met me and said they were all going to Bishnauth by Sunday steamer, so I shall only get one night there. We had tea with Mrs Godwin and in the middle Mrs Drake and Miss Gill came in and finally old 'Conscience' himself. He looked *muggra*, as usual, but was very cordial. When tea was over he and I strolled over to Phoolbari in order that I might send a wire to Felce and it was dusk when we got back. Walter dined with us and we closed with bridge.

Sat. Mar. 3rd

Godwin and I went down to the Thakurbari range as he wanted to shoot part of his course because he was not a marksman. He reshot at 500 yards and increased his total to 163 but could do nothing at 800 yards so we returned and went through his pruning, some of which I liked, especially his stick pruning. There were troops of monkeys in his plot near the jungle. I had never seen so many. Some were quite large, at least two feet at the shoulder when on all fours.

Said goodbye to the ladies and drove to Thakurbari. I was only just in time for the train to Borjuli and in the carriage were the Arthur Moores nurse and a man who proved to be Thompson, the extra assistant deputy commissioner. He was going to see Eric Hannay about a case and told me he had not been up the railway before. At Borjuli Station Clarke was waiting for me as Felce was away shooting; also the two Hannays were loafing around, evidently waiting for Thompson.

Called on the Eridges and reached Tarajuli just after dark. We were feeling very fresh and after dinner felt impelled to nautch to the accompaniment of his gramophone. It was the most amusing evening I have had for a long while.

Sun. Mar. 4th

Walked round the garden. There is splendid wood on most of the bushes. Ainsley and Scott Erskine rode in for a peg. Had dinner with Eridge. He gave us an A1 blow out. We left at midnight with many promises not to give him away at home.

Mon. Mar. 5th

Tarajuli garden is wonderfully organised, though Felce is constantly away and Clarke, though very smart, quite new to the work. Felce arrived about 6.30 p.m. in great spirits and made us go over and dine. He had seen lots of game but his only was a sambhur. We heard a lot of the scandal at the Palace. It was quite like Kipling.

106

Tues. Mar. 6th

Rose at 5 a.m. Then Clarke drove me to Rangapura for the trolley and saw me aboard. He is one of the best and I am very glad I stayed with him. I found both steamers at Tezpur and hadn't long to wait. Sent my bearer home. There were five passengers: the Nowgong sergeant-major, a Mrs Cleave, a brother of Fife, a most awful bounder, and another man whom I fancied was a government official. There were clouds of dust in the air and the view was poor; I could hardly see Singri Parbot before we reached the ghat. However, the air cleared before reaching Gauhati but the sun was too high to give the full effect and owing to the dry season Peacock Island was joined by the sand to the bank, so I was disappointed after the perfect view I had seen before.

The ghat lay right below the island but as I wanted tobacco and stationery I walked into the town with Fife and the sergeant-major and we went on the dak bungalow. The place brought back pleasant memories. We returned in a ticca gharry but owing to the sand we could not approach the steamer, so hearing her whistle, we thought we were late and ran like hell. But it was a false alarm, as she was merely drifting down to a lower flat.

Wed. Mar. 7th

Woke up to find we had stuck somewhere below Goalpara. However, just as a steamer came to our assistance we got off. The river here was not very interesting and about 11 a.m. we tied up at Dhubri. I went on shore with Fife to put his luggage on a floating dak bungalow, which was very nicely got up and where meals were served.

After we left Dhubri a regular sea got up. The waves must have been three feet high and there was a strong headwind. As it got dark and heavy shower of rain fell, but once over the wind dropped and the water calmed down.

Thurs. Mar. 8th

The river was again uninteresting until noon when we passed a corner where another large body of water joined us, very sandy in colour. For some distance there was a defined mark down the centre of the channel, as the waters were slow in mixing. I was told it was the Ganges. We reached Goalundo about 1 p.m. and I was surprised at the number of boats there.

Fri. Mar. 9th

I found the river much the same as at Goalundo and almost too broad to see across. We reached Chandpur at about 11 a.m. and then turned into a well-defined channel about 200-400 yards broad with banks not more than three or four feet above the water. The river evidently never cuts here and cannot ever rise as houses are close to the edge in many places. The country is absolutely different from that above Goalundo, as it is well populated. All the turf is closely grazed off and it resembles a sort of park. Here and there are groves of tamol trees and others whose names I do not know. There is evidently very little thatch as most houses are tin-roofed. There are no buffaloes and but few goats, though plenty of cattle.

Previously I had an idea that the sunderbunds began immediately below Goalundo, but the *serang* said it would be another day to reach them. The river pursued a very serpentine course and in some ways reminded me of the country near the Broads.

About 4 p.m. we arrived at a town lying on the west side and there were a number of steamers and flats. Also I saw what looked like a mill and further on was a church. I was told it was Barisal. It looked very pretty in the hot afternoon sun and from the ghat running south along the river bank was a fine shady road with some well-built houses standing in compounds. We dropped the flat we had brought from

Goalundo and at 6 p.m. we stopped at a small place to pick up bags of *supari*, grown, to my surprise, by a colony of Chinese. The energy they displayed in loading was in great contrast to the Bengalis.

Sat. Mar. 10th

Just after breakfast a network of channels commenced and we entered a narrower part with low tree jungle right down to the water. This was the sunderbunds. It was interesting and pretty. I got out a pistol but only got one shot at a muggar, which I missed. Then the channel broadened out and another boat approached. She hailed us and anchored and we took two passengers on board. They were a cleanshaven man, who might have been any age between twenty and forty and a very fat, flabby-looking woman who looked old enough to be his mother; but they slept in the same cabin so evidently have another relationship.

At 6 p.m. we turned into another very narrow channel not much more than two ships' breadths and as it is tidal here we had to wait, owing to lack of water. The moon rose and the scene was weirdly beautiful, yet with a grotesque element: ahead the narrow channel with primeval jungle right to the water's edge and all around unbroken silence. Yet, as a centrepiece, looking horribly out of place, was an ultra-new fabric trying with its searchlight to outvie the Queen of the Night, and now and again disturbing the solitude with discordant blares.

At dinner the newcomers discussed in undertones; the woman appears to be an American. Both of them seem to have travelled, as he mentioned South Africa and the Canaries. He asked her how she ate pineapples and explained our system of cuting them in halves and using a spoon. It seemed rather funny for a husband not to know already, but as she was not attractive and they brought two servants, I supposed they were tied up.

Sun. Mar. 11th

We were again in broader channels, average 300-400 yards. The *serang* says that owing to our delays by fog and tide we shall not reach Calcutta till late this evening. Cultivation has reappeared and in many places sea walls have been constructed and jungles cut down, but crops have evidently not done much yet. What jungle there is is much scrubbier than previously.

All the way from Goalundo I was struck by the number of country boats which appeared to do most of the traffic, whereas up above the steamers do it, the current probably being prohibitive.

3 p.m. and we were clear of the sunderbunds and right ahead was the first ocean-going boat I have seen in five and a half years – a one funnel, two masted cargo boat with high foc'sle, southward bound. 5 p.m. and we were running north up the Hooghly, which is very broad here. Passed three other boats at anchor. 9 p.m. and the *serang* said we were getting near Fulta. The channel has become narrower now.

Mon. Mar. 12th

Rose to find us anchored in midstream with Howrah Bridge about a mile upstream. Hailed a dinghy, a kutcha-looking boat pulled by two men with a third as a steersman, part of the stern being covered like a kyah's cart. Arrived at RSN Co. pontoon and got a ticca gharry. Drove to the Great Eastern and was given a room at the top storey. The heat was frightful so had a bath and got into white things. Had *chota* and walked to the office. All were very busy, it being mail day, so I cleared out after being introduced to Denne of Surma. Tiffin over, I did some shopping, including jobs for Dunlop, which put in time till I had had to change for dinner. The dining room and reading room have electric fans and are beautifully cool.

110

Tues. Mar. 13th

Went to the office again and met Hamilton of Central
Dooars, also Denne. After tiffin paid bills at Whiteways,
etc., got my ticket at the P & O office, where the boss said
he knew Harry well, and sent his regards. Took my luggage
to Kidderpur dock and fell in with my steward. He took
me on board and showed me my cabin, which is intended
for three but one man, Lord Kinsale, is not coming, so I
am in luck. Drove to the zoo and was much disappointed.
Saw a rhino and went to the reptile house, but only saw a
sick python, a Russels viper and some cobras. There were
a few others of less note but no hamadryads, which I had
understood were very fine here. After this it was too dark
to see much more so I drove back. There were crowds of
fine carriages about and a surprising number of motors.

Wed. Mar. 14th

Left the hotel about 7.30 a.m. At Chandpal Ghat were a
crowd of people and we had to wait for the medical
inspection which was commenced shortly after 8.30 a.m.
and we at once went onto the tender and cast off. In less
than half an hour we made fast to the *Sardinia* at Garden
Reach. After putting my things in the cabin I went on deck
and took stock of my shipmates. At first sight it struck me
that there were very few girls and none very attractive, that
the women largely outnumbered the men, that the majority
of the men were over thirty-five at least and that there was
a crowd of children. My stablemate is named Macdonald;
elderly, wears a pince-nez, is slightly built, has a rather dry
manner and a wife and child – a demure-looking, pretty
little girl about eight whose name is Maisie. I noticed Mrs
Sweet on board; she has taken the skipper in tow at once.
  They gave us breakfast and about 11 a.m. started getting
up anchor. Booked a seat at a corner table. Went on deck
and got in conversation with a young chap about my age

111

whose name is Young. He has tried jute for over a year and is giving it up. His people own shares in some gardens and his uncle is Judge Sale, who threw out the Bain case.

About 5 p.m. we passed the dreaded 'James and Mary Shoal', a quicksand capable of pulling anything down; in fact a big city boat was lost there some years ago. We went dead slow with an officer in the chains, heaving the lead. The shoal extends right across and is caused by the Roopnarain River coming on the right bank. About 7 p.m. we anchored for the night, and seeing the pilot had a yarn with him. He told me all the pilots were either Worcester or Conway boys trained up to this job quite early. Turned in at 11 p.m., at which time lights were put out, except two in the smoking room.

Thurs. Mar. 15th

We got rid of the pilot in the evening off the sand heads, which are really sand bars and not visible. Got to know the smoking habitués: Major Wolfe Murray, dark, slight and said to be the best player in the Bengal Club; Captain Ridgeway, short, medium weight, rather Irish; Lieutenant Walkertall, small, moustache; Kingsford, Chief Presidency Magistrate of Bengal, dark, rather good-looking; Drake Brockman, a big man, heavy moustache, apt to be didactic but very nice. All these are bridge players. Black, another big man, is in the tea firm of Shaw Wallace. Le Fanu a small man, oldish with a moustache and beard but no whiskers. I think he was a judge in Madras. There were several other men, not in any way remarkable.

The leading ladies were Mrs Drake Brockman, very handsome, tall, dark and rather a grande dame; Mrs Ridgeway, medium height and not bad-looking; and Mrs Wolfe Murray, who must have been a beauty once, but is now rather faded. The girls were Miss Burt, Miss Henderson and Miss Robertson, all nice girls but not remarkable.

Fri. Mar. 16th

Very hot in the cabin as the breeze is on the wrong side. Punkahs are only supplied on order at Rs. 15 and I neglected to make arrangements.

Sat. Mar. 17th

I am having bad luck at bridge. These men are streets above me, and I have lost Rs. 35.

Sun. Mar. 18th

Captain read prayers but I didn't attend. Young is running Miss Henderson and attempting Miss Burt but her mother watches her like a hawk. At midnight we were off Ceylon.

Mon. Mar. 19th

Woke up to find the coast quite plain with surf breaking on the shore. At 8 a.m. we passed Galle which, through a glass, looks pretty. At noon we sighted Mount Lavinia, which is a big hotel lying south of Colombo and distant about seven miles. Then Colombo itself appeared with a lot of shipping behind its breakwater, including a cruiser, the *Sutleg*. We got in and anchored about 3 p.m. But after the health officer came on board, two people were found with infectious diseases; one a lascar and the other a boy in the first saloon, who has got scarlet fever, so we weren't allowed on land. At night they started coaling and everywhere was shut up, so the heat was awful.

Tues. Mar. 20th

Left Colombo just after midday and all were glad to be on the move again. We held a committee meeting to choose members to organise deck sports, etc. I was made secretary and also responsible for the daily sweep.

Fri. Mar. 23rd

Four ladies have now won the sweep and no men. It looks as if the skipper was faking the returns.

Sat. Mar. 24th

To my surprise I won the sweep amounting to 28s. It was our biggest run: 301 miles. I thanked the skipper, who said he thought I deserved it.

Tues. Mar. 27th

The African coast came in view about 10 a.m; a high coastline of sand and rock. It was a pleasure to see land again. During the evening we passed Aden but didn't call.

Wed. Mar. 28th

During the early morning we passed a number of islands with a pink mist over them. To me they looked very beautiful with white surf breaking on the beach.

Sat. Mar. 31st

The evening was much cooler and I put on a thick waistcoat.

Sun. Apr. 1st

Woke up to find us at Suez. The health officers came on
board and were inspected by a woman. It was a farce, as
usual, as she merely ticked off our names. I discarded my
white things for good and got into tweeds. About noon
the *Mongolia* from Australia arrived. She appeared very
full up. At 1 p.m. we entered the Canal and after breakfast
I sat on deck and read. My tinted glasses came in handy,
as the glare from the sand was severe. About 4 p.m. we
entered the Bitter Lakes, and after dinner left Lake
Timsah.

Mon. Apr. 2nd

Arrived at Port Said, 6 a.m. and started coaling. Rose about
7.30 a.m. and felt a chill in the atmosphere at once and
was quite glad of my cloth clothes. Went ashore, met
Walker and went to a second class show called the Con-
tinental. Had breakfast and then looked in at a music hall,
the Eldorado, where we found Black and Young. Only
stayed a few moments, then loafed for an hour or so and
returned, not having effected much. Boat left about 10.30
a.m. and when clear of land a heavy sea got up. About tea
time the seas were breaking over the foc'sle. The *Osiris*
passed us and appeared to be having a very poor time.
Every now and then her bridge was covered with spray.
After tea took Miss Burt to a sheltered place by the bridge
to watch the waves, but her mother found us and took her
away.

Tues. Apr. 3rd

We only made 196 miles in a very poor run. Sea still high
and weather bitterly cold.

Wed. Apr. 4th

Much calmer as we passed close to Crete. Still feeling seedy and have decided to get off at Marseilles.

Fri. Apr. 6th

Passed the Straits of Messina, but the view was not good. Ran under Stromboli but there was not much to see, owing to a cloud of steam hanging over the crater.

Sat. Apr. 7th

Woke to find us running up the coast of Sardinia; a rugged, rocky coast with but few buildings and nothing of interest to see. After breakfast passed the Straits of Bonifacio. Held my last sweep, which was won by Dr Hawkins. During the afternoon we had the gramophone and I sat and talked to Miss Burt. She interests me considerably and we went down and had tea together. After dinner I didn't stop for coffee in the smoking room as usual, but made tracks for the companion to intercept Miss Burt, which I succeeded in doing, and we had a very pleasant evening. I wish I had got to know her earlier in the voyage. Her great failing is that she is not musical and doesn't like poetry.

Sun. Apr. 8th

Rose to find us in the outer harbour of Marseilles, with the Château d'If close at hand. We had to muster just before breakfast in the saloon for the health officer's inspection. He was horribly fussy and would not start till every man and woman was present, and as Mrs Wolfe Murray was ill it was some time before she put in an appearance. As usual the inspection was a formality for

116

we merely answered our names, and when it was over the gong was rung, for which we were devoutly thankful. Breakfast over, I tipped my stewards: cabin steward £1, table steward 10s., desk steward 10s., bath boy 2s.

About noon we entered the docks and tied up, so I went on shore and bought some violets. I had intended having lunch onshore with Black and Young but it was nearly 1 p.m. before I got my baggage and ticket given to Cook's man for conveyance to the station with Stiefelhagen for the 8.20, so I stayed on board.

Lunch over, I found lots of people saying adieu but I didn't bother about them much, excepting the Burts. Black, Young, Stiefelhagen and I drove into the town. When fairly among the streets we got out and walked to the zoo, the entrance of which was very fine, but nothing much to see inside. It was a trying climb, as we started with a flight of a hundred steps or more. What impressed us most was the grass; it looked lovely and I could have rolled in it.

We walked back to the ship, had a wash and drove off once more for dinner at the Basso restaurant. We had an excellent dinner and mopped up three bottles of Chablis which, combined with the excitement, had a wonderfully exhilarating effect. We almost embraced the garçon and interpreter and then tumbled into a carriage and were driven to the Gare du Nord. There wasn't much time to spare and after drifting about like a flock of sheep I saw Cook's man standing on the platform and Stiefelhagen looking out of a carriage above him. I climbed in and got my luggage all right. Paid Cook's man his bill of 5s., said goodbye to Black and Young and the next moment the train was off. Five minutes later and I would have missed it. The train was a corridor only, with the alley way down the side, not through the centre. In our compartment were the Stiefelhagens, child, ayah, myself and one other man. I had no greatcoat to make a pillow and so didn't sleep very comfortably; in fact only dozed.

Mon. Apr. 9th

Day seemed to dawn very early and it was chilly. We stopped at Lyons for a few minutes and we got some rolls. We reached Paris at 11 a.m. and said goodbye to the Stiefelhagen family. At 11.30 a.m. got a train going round the circle and landed myself at the Gare du Nord; only had about fifteen minutes to wait and was delighted to find English-speaking people in my carriage. A young chap sitting on my right was very talkative. He wore long hair and seemed to be a commercial. Said he came from Biarritz, spoke of 'His Majesty' and remarked that he had seen him looking at a villa a few days before. In front of me was an elderly, stout person and her maid, who looked the more attractive of the two. The stout one tried to unhook her stays from the outside, but apparently without success.

There was no trouble with the Customs at Calais and we went straight on board. The boat was turbine-driven, the first of the class that I have been on. We left in a few minutes and were quickly clear of the harbour. It was slightly rough and one or two ladies were sick. It was misty and I couldn't see the English coast until we almost got to Dover pier.

What a blessing it was to be on English soil again and be able to speak one's own language to everybody! I enjoyed the train journey, as it was a fine evening, but when we got out at Victoria it was cold and I shook as if with ague all the way to the Arundel Hotel. The first thing I did there was to get in a steaming hot bath in which I lay and soaked. There was a fire in the bathroom, which made things perfect, and I dressed in front of it.

I had a very fair dinner with coffee liqueur and a 6d. J.S. Muria cigar to follow. Sixpence seemed a lot for a cigar to a man accustomed to give 4s. per hundred. Bed, including bath, table d'hote, breakfast and attendance cost 5s. per day.

Tues. Apr. 10th

Took a hansom and drove straight to the hatter recommended by Davis, where I purchased a topper, bowler, Monte Carlo and cap, also a new type of hatbox holding two: total cost £3. Then to the tailor at No.1 Bishopsgate Street Without, as the name has been changed, I had some difficulty at first, but at last a bobby put me on the scent. I ordered a frockcoat, etc., and a lounge suit, both to be made correctly, whatever that means nowadays; also a greatcoat and two spare waistcoats, one white and the other fancy. The people were very obliging, probably as I mentioned Davis's name. Bill will work out about £15.

Next I went to the office and, on inquiring for Russel, was told they had a board meeting. However, I sent in my card and Pat appeared and shook hands in a very friendly way and asked me to come the following day at noon. After lunch I went on a 'bus' towards the west to look at some motor shops, as Young had been extolling the excellence of tri cars. I had never heard of them before but believe they would do well for Assam. I saw one but it had a bicycle seat for the driver so I didn't care for it.

Wed. Apr. 11th

Got a pair of gloves in the new shade, a sort of slate colour, and got to the office by 11 a.m. Russel turned up and asked a good many questions; also told me the sales had worked out about 8d. against my estimate of 7.90d. and Dunlop's 7½d. The profit should be well over £3,000. Russel is rather keen on monorails and thinks we could use one at Monabari. He has given me a pamphlet to go into the question with and make a report.

On leaving him I went to have my things fitted and then back for breakfast. Left for Waterloo and caught a fast train for Bournemouth Central. The place was very familiar and I didn't feel as if I had been away for nearly six years.

119

At Studley Tower I found Mother at the door, looking just the same, with Dad behind her. Mary had gone to the Central but was the only one who had changed her hair, having gone very grey.

We had a very jolly dinner and I felt as if I had never left the country, or anyway for only a short time. Everything was just the same, including the evening prayers at 9.45 p.m.

Burra Bungalow, polo ponies and buggy.

F Troop Assam Valley Light Horse on parade.

Pruning tea bushes.

Planters and wives. Writer of Diary cross-legged on right.

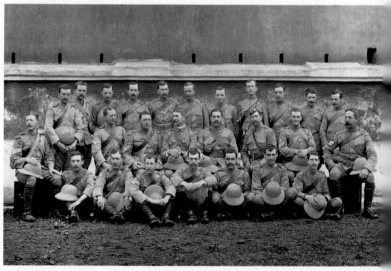

F. Troop AVLH

Apr 10 — Spr 27 → 5½ months

7

MY SECOND TOUR OF INDIA

Thurs. Sept. 27th

Spent the morning packing and at 1.15 p.m. drove to the station. We reached the Arundel at 5 p.m., had tea, smoked and went to bed, before which I wrote to Agnes.

Fri. Sept. 28th

After breakfast went to the City, approved and paid for my flannels and visited the BI office, but they had no news of the car. Called at the P & O office at Lime Street where I said goodbye to Russel. After lunch called at the BI office again and heard that the car had reached the dock. After dinner Ed (Rev. E.C. Hetherington) went to call on the Bishop of London and I talked to Father on my most interesting subject. When Ed came back the parents went to bed and Ed and I talked till nearly midnight.

Sat. Sept. 29th

Got a letter from her, which improved the outlook. Said goodbye in my bedroom. It was rather an ordeal and does not require to be dwelt on. My friend the hall porter put my luggage in a hansom and Ed and I drove off. Arriving at Victoria, got my luggage weighed and paid 10s. excess.

I booked a seat also in the train from Calais. Left at 11 a.m. and at Dover got the new turbine boat, *Invicta*. There were two Japs in the carriage and a man who turned out to be in Ridgeway's regiment. We shared a cab to the PLM. It was a taximeter, the first that I have seen. Had a good dinner together and left at 9.20 p.m.

Sun. Sept. 30th

Got coffee at Valence and reached Marseilles about 10 a.m. Half an hour late. Drove down, discovering on my arrival that the *Egypt* was lying alongside the quay and I had to take a rowing boat to the *Syria*. We left at noon and at lunch I found I was at the doctor's table: a Scot with an accent and no sense of humour. After dinner, lounged on deck. The moonlight was glorious on the water and I could have sung 'Come o'er the moonlight sea' with great fervour.

Mon. Oct. 1st

My table consists of the doctor and opposite Mrs Challoner, a gay widow who is fond of cigarettes and liqueur brandy; Lawrie, a planter, Miss Iles, a quiet, ladylike girl going out to be married; Duguid, a red-haired Scot civil engineer; Mrs Lyall, small and frivolous, a pal of Mrs Challoner and Mrs Compton, middle-aged and motherly. In the smoking room the best men are Major Herbert, a Civil Service man, Horncastle and Garlick. In the second class I met Stevenson and Percy Forbes. Ferguson is also a good chap.

Tues. Oct. 2nd

Held a meeting to choose a committee. Allinson is secretary, Major Herbert, Horncastle, Ferguson, McPherson and I

being the rest. Horncastle and I arranged an impromptu
dance which was very successful. I am also playing bridge
in Allinson's set and doing fairly well. Our points are 10d.,
which is mild.

Fri. Oct. 5th

We reached Port Said at 5 p.m. and I invited the Allinsons
and Miss Wetheral to dine. Percy Forbes had sent me a
letter earlier in the day, inviting himself to dine, but I
declined his kind request for which I have no doubt he
does not bless me. Port Said seemed to smell worse than
ever. We strolled round the front. Soon getting tired of
this, we went to a hotel for dinner but as they couldn't
serve us till 8 p.m. we returned to the ship. We left at 11
p.m. and in consequence of a brilliant moon and our
searchlight, the Canal looked weirdly beautiful.

Sat. Oct. 6th

We had to tie up during the night and so didn't pass
Ismailia till 10 a.m. Lake Timsah wasn't as big as I had
expected, never having seen it by daylight, but Ismailia
made a pretty patch of green against the sand. Our passage
of the Canal was very slow and we didn't reach Suez till 6
p.m. and left again about 10 a.m.

Sun. Oct. 7th

An off day and in company with most men I didn't play
bridge. There was no service in the first class but one at 8
p.m. in the second, which I didn't attend; we sang hymns
instead.

Thurs. Oct. 11th

Reached Aden about 10 a.m. and a good many went on shore but I thought it too hot. I am troubled every day with small headaches which are annoying. There were no diving boys as some were killed by sharks and the government stopped the rest.

Sat. Oct. 13th

Arranged a mixed concert with the second class, to be held on their deck. Our performers was Miss Iles, Miss Stead, Mrs Ferguson, Miss Allinson, Robertson and Horncastle; on their side, amongst others, Adams and the mad anarchist woman who was ill in her bunk till the evening, when, on hearing about the concert, she got up and insisted on taking part. On the whole it went off very well.

Tues. Oct. 16th

Finish of the bridge tournament. Allinson led for a long time; then I passed him, only to be beaten by Horncastle.

Thurs. Oct. 18th

Nothing to enter except there were games after dinner, such as Nuts in May, which I didn't join in.

Fri. Oct. 19th

Got inside Colombo breakwater about 10 a.m. The diving boys were in great form and all the newcomers, especially Miss Iles, were much interested. There were a good number of ships including a Japanese cruiser. I arranged to go on shore with the Allinson party and as the skipper said we couldn't leave before Saturday 8 p.m. we decided to go to Kandy. Arriving at the pier in the PO tender about 11 a.m. we took a gharry and drove to a shop where the ladies looked at some silver goods, then along the sea front to the Galle Face hotel ordered lunch for 12.30 p.m. and amused ourselves in the spare time having drinks and then looking at silverware, Japanese cloisoni work and jewellery. Mrs Allinson bought a Japanese silver vase which was beautifully executed but I took nothing though much tempted by a blue moonstone necklace. The dining room was a splendid apartment and fully thirty feet from floor to ceiling, on one side overlooking the sea. At 1.30 p.m. we left for the station in rickshaws. I had never been in one before and thought it very comfortable. After getting a good carriage settled ourselves in and ordered tea. We left at 2.10 p.m. and on leaving the town ran through low lying country principally dhan kets and covered with water. I felt strangely as if I was coming home and absolutely delighted at watching familiar scenes, certainly the east has a powerful attraction for me. This part of the country is very like Assam. At 4 p.m. we stopped at a station and our tea was handed in, it was better than any we have had on board, they also gave us rolls, butter and jam. On leaving this station we got into more hilly country and the railway ran round the face of the hills. In some places there must have been a drop of 1,000 feet at least, the slopes were terraced and dhan was being grown.

It now began to rain and once or twice we were able to look down on a cloud, at another time, the clouds seemed to rest on the hill tops above us just like a curtain or, rather, ceiling cloth. The view over the valleys was grand and never for two minutes the same owing to the incessant curves of the track. We passed several tea gardens but the bushes were small and meagre and I noticed that many more trees of other descriptions were grown among the tea than is our custom. It was dark when we reached Kandy at 6 p.m. and raining fast but we got a gharry under cover and in a few minutes were at the Queens Hotel; a good sized place, very comfortable and not many people were staying there. Dinner at 7.30. A better meal than for many a day and afterwards had our coffee and liqueurs on the verandah; then seeing it had cleared up we walked round the lake or tank.

Sat. Oct. 20th

After *chota* we took rickshaws and visited the Buddhist temple. I had never seen one before and found it very interesting. First of all we threw rice to the sacred tortoises, then inspected the audience hall of the Kandyan kings, now turned into law courts, the teak pillars of which were beautifully carved. The temple itself had a number of shrines with buddhas in all kinds of metals and ivory and in one place was enshrined the sacred tooth, but we only saw the casket. I should have liked several hours there but we did not have the time.

Then we went up a hill through beautiful tropical vegetation, passing tamol, cocoa, nutmeg and coconut trees; also breadfruit and fine clumps of bamboo. When near the top we got out and finished the journey on foot to a spot where we could look down on Kandy with its lake, roads and streets laid out like a map beneath us and its green, blue and purple hills hemming it in on every side. We retraced our steps to the rickshaws and came down a different way, past Government House. Our train left at 2 p.m. It rained most of the way, so we were very lucky in having had a fine morning. Got aboard at 6.30 p.m. and left at 9 p.m. The P & O superintendant came on board and said he had met Harry, who had asked him to look out for me.

Wed. Oct. 24th

Arranged a big dance for the evening with programmes

and all correct. About the middle of it we reached the pilot brig at sandheads, so dancing was delayed till the pilot came on board, when we heard that owing to neap tides it would take us till the following night to reach Diamond Harbour. The news was annoying, especially to Miss Iles, who had hoped to get married tomorrow. Dance continued till 11 p.m., when Robertson and I were cheered and 'Auld Lang Syne' sung.

Thurs. Oct. 25th

We lay at anchor till 11 a.m. and then slowly moved off. At 2.30 p.m. passed Saugor Island and at 6 p.m. anchored in Diamond Harbour. Some went on shore to go by train but Miss Iles remained; also her captain, who is a very plain person.

Fri. Oct. 26th

Breakfasted at 8 a.m. and at 9 a.m. the tender left. The journey was uninteresting, except to mark down the James and Mary Shoal. We reached Chandpal ghat at 2.15 p.m. and everyone was famished. I drove to the Grand and got a decent room with bathroom attached and a fan. Washed and had the remains of a tiffin. Major Herbert came over and suggested I should sit with him and his wife in future, which invitation I accepted. I had tea in my room at 4.30 p.m., wrote letters and dined at 8 p.m. with the Herberts and Napier.

Sat. Oct. 27th

Went to the office and collected a letter from Godwin about factory work. Was unable to see Bowery and was told I must stay till Monday as the question of my agreement was to be discussed.

Mon. Oct. 29th

Went to Marshall's and saw among other objects of interest a new style of boiler and a patent spreader for paragons; also to David's for various things; lastly to the office, where Bowery told me the directors are offering me a three year agreement at Rs. 300 per month, which is not good enough, but I shall have to talk the matter over with Dunlop. Decided to leave for Goalundo tomorrow at 7 a.m.

Tues. Oct. 30th

Wet morning, so the cyclone is still hanging around. Had *chota* at 6 a.m. and left at 6.30 a.m. Reached Sealdah in about ten minutes and got a carriage with De Tivoli. The country all the way to Goalundo was very wet and there was evidence of flooding. Our steamer, the *Wazari*, left about 12.30 in a rainstorm and there was no view.

Wed. Oct. 31

Another wet morning.

Thurs. Nov. 1st

Woke up at Dhubri where we waited for the train till about 10 a.m. Only one man came on board. Christie, the manager of Cinnatollah, just out from home.

Fri. Nov. 2nd

Went on deck to find us nearing Gauhati and presently Peacock Island came in view, this time separated from the mainland; the ghat is in the old place. Went on shore and sent a wire to Dunlop and discussed accumulators with the office man. I hope to get mine charged there. Left Gauhati at noon and at dark were just beyond Mangaldai Ghat; there was a magnificent sunset. Said goodbye to De Tivoli. He has arranged to supply me with petrol.

Sat. Nov. 3rd

Left Tezpur at 7 a.m. The Nowgong sergeant-major came on board, but no one else. The river has altered considerably here. We were two hours late reaching Silghat and wasted another there. Breakfasted just before Bishnauth Ghat which is in a good position under the *busti*. We wasted more time here as we had to shift the flat 100 yards downstream. Left Bishnauth at 2.20 p.m. and very soon the old familiar clump of jungle in Kationabari appeared, but there is a mile of sand in between. At 4 p.m. tied up at Behalimukh and found Radhoo waiting, but no buggy. This turned up, however, in about quarter of an hour with Anandi syce and Burton's pony.

The road was awful and I had the buggy led to the bridge at the ghat; there had evidently been heavy rain. After this I got on faster but it was dark at the river where the sand was a bit soft. When I passed up the tea house road a voice which proved to be Punia's asked how I was; it was a cheery greeting. At the bungalow were Dunlop and Burton and we were all glad to meet again, but Dunlop is a bit seedy. There was a lot of news, including Glass's engagement to Miss Scott. Koch has been sacked from Belsiri and Cowper appointed. Our crop is about 1500 maunds ahead. Davies has applied for leave, having been seedy, etc. I dined and also slept at the burra bungalow.

Everything has gone like clockwork, which is very satisfactory.

Sun. Nov. 4th

Went to the office and tea house before *chota*. Everything including the babus, as usual. About 10 a.m. we left for the polo ground as a cricket practice was being held. There was a grand view of the snows and I much enjoyed seeing them again. At the ground were nine men who ragged me a great deal but all to no purpose, as I kept silent. Played bridge after lunch and left about 5 p.m. I again dined at the burra bungalow as my mosquito curtain hadn't arrived. I had a big dak to read and her letters were most satisfactory.

Mon. Nov. 5th

Walked round the new line, including the brickyard on the Gingia road. Visited the nursery and new drain, which was my notion; then down to Kationabari, where my nursery has not turned out well.

Tues. Nov. 6th

Wrote to the agent accepting their agreement but asking for an increased screw. Dr Smith came to see Dunlop.

Wed. Nov. 7th

Went out shooting with Dunlop at Kationabari. The usual round, as we both shot badly and only got a leash of partridges and a jungli murghi, though we killed another brace of birds which weren't picked up. During the afternoon we went to polo. There were only five players but Adams was allowed to knock a ball about for practice afterwards. Everton left early.

Thurs. Nov. 8th

As the lieutenant governor and his retinue were to visit the garden on Friday a muster of coolies had been arranged and all the lines cleaned. During the evening Dr Smith, Lawes and Davis came over.

Fri. Nov. 9th

A pouring wet morning. It began to clear at 8 a.m. and I biked to Kationabari to march out coolies. Returned to Monabari and changed into store clothes, then went to the office and got the coolies mustered according to their jats. It was now a fine day but no sign of the LG. At 10 a.m. went to the burra bungalow where Dunlop had just received a note from Thomas, our DC, saying he was not coming. This, considering our trouble, is an absolute insult. There was nothing, however, to be done so we broached the 'fizz' and smoked the special cigars. Johnny and Duguid arrived about 12 a.m. and then went off to lunch on the Brahamakund. They returned about 6 p.m. and said it was Thomas's fault and the LG had expressed regret and was coming on Saturday instead.

Sat. Nov. 10th

At 8.30 a.m. got the coolies lined up and at 9.15 a.m. the buggies were seen approaching. First came Thomas (who is practically a babu) and Denny on horseback, then Hare and Mrs Lyon in one buggy and Monahan and Lyon in another. They stopped near the leaf house and shook

131

hands all round. Dunlop escorted Hare and I followed with the Lyons. Mrs Lyon had a camera and took several pictures, including one of the Allahabad women. We then walked round the tea house and back to the bungalow. Here we found Denny, who said Dunlop and Hare had gone for a drive round. They turned up a few minutes later and all partook of some light refreshment but refused simpkin. Immediately after they took their leave and the great event was over. All the coolies went to pluck. Dunlop got a wire from Wright, the man from New Glencoe who is coming here, saying Arrive Sunday. Filkin stayed the night and Captain Mears arrived.

Sun. Nov. 11th

Dunlop drove Mears, Filkin and me with the Bishop. The latter is a lazy moke and bores over to the right. There was a good muster on parade and we had carbine drill followed by mounted work, which I didn't engage in. Then tiffin, after which the adjutant lectured us on field reports and we finished up with polo. All adjourned about 6 p.m. and on reaching home Dunlop and I started playing billiards, in the middle of which Wright was escorted in by Burton. He is very tall, about six feet four inches, but looks lanky as fever has pulled him down; but the main thing is, he is a gentleman.

Mon. Nov. 12th

Dunlop sent men to fell timber at Bargang and took Burton over to breakfast there. I inspected the pathar with Wright and tired him out with the walk. He told me he had tried three times for Sandhurst and failed. He had done three years motor engineering and been six months in a bank. The worst thing is that he knows nothing of steam so I shall have to do the machinery this year instead of pruning.

132

Wed. Nov. 14th

Dunlop drove me to polo but only two a side turned up so there was no game. We drove Yerraman home and Dunlop being, as usual, careless after polo, smashed the railing off one of the bridges because the hub of the wheel struck it a heavy blow.

Fri. Nov. 16th

Danter put me through my drill, having stayed the night. Dunlop was very funny during the evening and lectured me on my want of affection for animals and, incidentally, human beings.

Wed. Nov. 21st

Dunlop took Burton for his farewell at the polo ground and Davis returned for dinner and the night.

Thurs. Nov. 22nd

Dunlop set Burton to make up box, account, etc., and when I came in to breakfast I found Wright sitting on our verandah with a dose of fever. Took his temperature – 105° – and put him to bed. At 1.15 p.m. Burton said goodbye and rode his pony to the ghat; he is taking it down with him. He was sorry to go in some ways but I think it will be good for him. At 6 p.m. Wright's temperature was normal and we dined together.

Mon. Nov. 26th

Building work, etc. Leaf has fallen to thirty two maunds.

It is the quickest close to a season I have ever known.

Thurs. Nov. 29th

Started hunting suitable ground for a ten maund nursery for Central Dooars seed. Decided on a place near the Madural Bheel and went there with Dunlop and some new line men, one of whom put up the old buffalo who lives there, but unfortunately he made off before I saw him. However, on the way back a wild boar rushed across the path. It was the first one I have ever seen.

Sun. Dec. 2nd

Went to see work near old brickfield at Kationabari and paid ticca. Returned at 12.40 p.m. to see Dunlop before he left for Tezpur races and received orders.

Mon. Dec. 3rd

Restarted plucking; thirty maunds.

Wed. Dec. 5th

Wright seedy again. Dr Smith came over, luckily, and saw him. He thinks the fever is working its way out of his system.

Thurs. Dec. 6th

Having heard of the arrival of the motor, sent to see if it could be put on a cart, but it was impossible, and the crate must be opened. Last days plucking a total of thirty maunds.

Fri. Dec. 7th

Sent two *mistries* to the ghat, also Minchand with some lunch, and then biked over myself. I found the crate on its side and we had some trouble in turning it over but none in opening it. The steering wheel was broken in transit, which is a nuisance. I got the car eventually on a gharry, then lunched and reached the garden at 3 p.m.

Mon. Dec. 10th

Dunlop went round everywhere and was satisfied.

Wed. Dec. 12th

Dry batteries arrived at noon and after breakfast we got the car going. Went to the burra bungalow and then started for polo. The road was worse than it looked and we had to go on our second gear. We reached the polo ground and found Davis with the Robinsons. After a little while Swinley arrived with Miss Gibb. Then Edwards and Butcher, all looking as usual. Wright and I left soon after the polo was over and had a good journey back till, within a few yards of the brickfield, the petrol pipe burst and we had to push the car home.

Sat. Dec. 15th

Wright again seedy. During the afternoon I put my big sprocket wheel on the car and ran it down to Kationabari. The coolies crowded round and it was amusing to see them. I returned on the third speed, which nearly shook the machine to pieces. Afterwards, going as far as the seed garden, I ran out of petrol and had to get it pushed back.

Sun. Dec. 23rd

Wright and I went for a shoot round Kationabari, but birds were not plentiful and we only got one. We tried a game of tennis during the afternoon but Wright was not on form.

Mon. Dec. 24th

Started Bijoy pruning. General work fair.

Tues. Dec. 25th

Christmas Day about the dullest I have known in this country. A very cold morning, down to 44° with slight fog. Work as usual and a crowd of absentees whom I interviewed at 5 p.m. Things generally unsatisfactory.

Sun. Dec. 26th

Wright rode over to Crutwell's and after getting a dose of fever didn't return. Dunlop and I worked in the office and drove to Gingia about 4 p.m. It was my first appearance and the place is changed as part of the compound has been abandoned. We played bridge and lost a good deal. Arrived home at 1 a.m.

Wed. Jan. 2nd, 1907

Wright drove me to polo. His pony is a very fast one in the buggy. There were few people present and we returned early.

Sat. Jan. 5th

Wright went to stay the night with Swinley for cricket
practice on Sunday. I dined with Dunlop, who was in an
affectionate mood.

Sun. Jan. 13th

Dunlop and Johnny left for Borpukhri and Wright for
Majulighur to attend the cricket and polo match v. Tezpur.
Ours is a mixed team, Bishnauth and Nowgong. During
the afternoon I walked down to the government road.

Tues. Jan. 15th

A fine threequarter eclipse of the sun was visible about
midday. I have never seen one before and so was much
interested. The daylight was dulled, though hardly so
much as I had expected and it was impossible to look at
the sun except through a smoked glass, Minchand of
course having prepared one.

Wed. Jan. 16th

Wright returned about 5 p.m. and Dunlop and Dey shortly
after. Bishnauth won the cricket and Tezpur the polo.

Thurs. Jan. 17th

Dunlop went round with me and was satisfied.

Sun. Jan. 20th

Wright drove Byron to Bargang while Dunlop and I rode down the Charali road, inspecting ground for garden extensions and also jute planting. About 4 p.m. we drove to Gingia for bridge and dinner.

Tues. Jan. 22nd

The Mozandara elephant came over and I went out on it to search for absconded coolies, accompanied by Baromaghi. We followed the edge of the *bheel* and went as far as the Gingia *nadi*, which does not fall into the haunted *bheel* but into one lower down, crossed the *nadi* and returned to the government road, turned south again and descended into the *bheel*, where I lunched at 2 p.m. After this, made an attempt to cross the *bheel* but the mahout (he was really a grass cutter) funked it, so we returned. Reached the factory at 4.30 p.m. having effected nothing.

Wed. Jan. 30th

Started drawing boiler tubes, which gave no trouble. Weather broke during the afternoon and at 6 p.m. it was raining heavily. Wright had fever.

Wed. Jan. 30th

Dunlop and Wright went to polo and Wright played for the first time.

Sun. Feb. 3rd

Dunlop worked for a while in the office and then went to

the Crutwells' opening breakfast. I spent the afternoon making a plan of the factory buildings for insurance.

Tues. Feb. 5th

Wright had another go of fever. Dr Smith injected him with quinine. The weather has now cleared up but we have had about four inches of rain, which is absolutely a record.

Wed. Feb. 6th

There was quite a sharp earthquake about 4.30 a.m., but no damage was done. As Dey and Davis were coming to tiffin I put the bungalow line women to prune my clearance. The two Ds arrived at 11.30 a.m. and Johnny went to the burra bungalow to change while Davis came to ours and I had the opportunity of some private talk, telling him of my future plans and receiving his congratulations and envy. They and Wright left directly after tiffin for the ghat on their way to Tezpur for the cricket match.

Sat. Feb. 9th

Went to the new lines to burn round the nursery. I had the Mohuris and Kacharis and a batch of women, but with even this number it was ticklish work, owing to the wind constantly veering. We burnt the side next to the Maduri *bheel* first, but when we came to the south side the fire raced away, so that we had hardly time to get to the small nursery and burn its edges before the main body of flame came upon us.

   This finished, we ran to the nursery by the Gingia road, where we found Chutar sirdar and some hoeing men and immediately started burning a fire line. By this time a line of flame about a quarter mile long was tearing through

139

the grass, making its own wind current, while the south west wind was blowing our newly-lit blaze in its direction, and when they met there was the most extraordinary result. The two bodies of flame rushed together, whirling round like a maelstrom, and were suddenly drawn upwards with a sucking sound into a regular wind, or rather, smoke spout. This, after a few seconds, came rolling over the burnt jungle towards us and I prepared to bolt, but it changed direction and presently dispersed. During all the years I have been at Monabari I have never seen anything like it before. There was now quite a lot of *shikar* going on and four or five young pigs, having taken refuge in the big drain, were caught and killed. A good day's work had been done.

Mon. Feb. 11th

Work as usual. Got wires from Wright, Johnny and Davis and they all arrived with Dunlop about 7.15 p.m. Dunlop's polo team was beaten in Nowgong. The cricket was drawn and Wright did not play in the second innings.

Thurs. Feb. 14th

Wright down with fever. Johnny and Davis came to tennis.

Fri. Feb. 15th

Rode Lettice to the range to start shooting my course. Called in at Butcher's and found he was going also. On arrival we found Robinson waiting but no one else turned up. Shooting was moderate and I made 98 against 100 last year, Robinson making 102. After finishing at 500 yards we stopped and went to Kettla for a drink. I breakfasted with Butcher and walked round his place, reaching home about 5 p.m. Wright was again ill and Dunlop had paid Kationabari.

Mon. Feb. 18th

Kolais Gousti absconded with ten men and five women, all good workers, beside several Faltu people. Dr Smith had breakfast with us and Wright went back with him.

Mon. Feb. 25th

Went round with Dunlop. Wright returned at breakfast time. He has fever badly.

Tues. Feb. 26th

Dunlop and I went to Kationabari. He has decided that I am to give all orders in future.

Wed. Feb. 27th

Wright drove me to polo and just after we arrived a tremendous hailstorm came on, the stones nearly an inch round. The storm lasted about half an hour and when over the ground was white, as if there had been a fall of snow. Polo was out of the question, so we returned early. Smith examined Wright this morning and has recommended his going home for the rains, as his blood is full of malaria.

Mon. Mar. 4th

As it was Davis last night he and Johnny came over and we had the rowdiest evening I have ever known at Monabari. I went to bed about 1 a.m. and shortly afterwards a tremendous storm burst, the wind a perfect hurricane. An inch of rain fell.

141

Fri. Mar. 8th

Pruning finished, also pay and pruning knives collected.

Sat. Mar. 9th

All hands on the hoe. Wire arrived from Calcutta, granting Wright leave.

Sun. Mar. 10th

The policeman, Halliday, from Tezpur, arrived in the afternoon. About my height with biggish moustache; seems a decent sort.

Mon. Mar. 11th

Halliday and Dunlop went to Bargang. I went out on the former's *hathi* and inspected the land across the river to see if we could cut a drain, but I fancy it will be a big job.

Wed. Mar. 13th

My new coil arrived by post, so instead of going to polo I fixed the car up and went for a run.

Sun. Mar. 24th

Swinley breakfasted with Dunlop. I went up during the afternoon and we all went for a drive around Kationabari. During the evening played billiards and had dinner.

Tues. Mar. 26th

Wright left at 7.30 a.m. and somehow I don't think his absence will be felt much. I ran the car and at last managed to get it to go on its top speed.

Sat. Mar. 30th

Dunlop went down country and I left at 5 p.m. for Gingia in the motor. It ran well and in spite of the road being newly thrown up I got over in half an hour. Johnny was in the bungalow and suggested going over to Crutwell's, which we did, and as the road was good the pace was likewise. We stayed with him till the moon rose and then returned to dinner. The car has now quite vindicated itself. I stayed the night.

Sun. Mar. 31st

After *chota*, went with Johnny through the forest to his jute cultivation, where a crowd were hoeing, and then to the river to fish. The distance from the bungalow was about two miles. I had never before seen the Bargang above the crossing it was very stony and the lights and shades on the water were exquisite, golden brown fading to yellowish green, then a patch of deep bluish purple where the current made a miniature rapid – a foot deep – over the pebbles. The banks on one side are forest with the trees in many places tumbling into the water and on the other side a sand bank with a sparkling, blinding glare rising from it, beyond which the sand grass jungle, of a lighter green than the forest.

We put our rods together and waded in (I having removed boots and socks) but the fish were shy and I didn't get a bite. I enjoyed the paddling where there was sand but the stones were decidedly painful and when I got out

of the water the sandbank scorched my feet, so my move-
ments were limited. Chishom's *pilkhana* were just below
our tiffin place and there were several *hathis* in the water,
squirting water over themselves and trumpeting now and
then.

After tiffin we walked back, had tea, and then I drove
Johnny back to Monabari in the car, but Dunlop hadn't
returned so he went back later in his own buggy, which
had been sent over.

Mon. Apr. 1st

Dunlop and I went to the jute land and settled a place for
the Nepali *chowdikar*'s house. While we were there we
spotted three jungly buffalo in a clear place just east of
the new brick kilns. Johnny came over in the evening and
I told Dunlop of my intention to get married this autumn.
This is the first time I have been without a fire and I also
wore white things for the first time this year.

Tues. Apr. 2nd

Went to see work on the jungle drain for turning the course
of the Bargang River. It will not be such a big job as we
thought at first.

Thurs. Apr. 4th

The new boiler tubes arrived and I started expanding.

Sat. Apr. 5th

Still raining in the early morning but cleared up 11.30 a.m.
when Dunlop and I went to see the drain. We had hoped

to see it finished enough for water to be let through but were disappointed. However, the sirdar reported in the evening that they had got water half way but it silted up fast and we shall need a flood to make it work properly.

Sun. Apr. 7th

Buddha and Moolla came in to arrange about opening a *busti* west of the burra bungalow, so I went and marked out the land for them. At 3 p.m. Dunlop and I left for Gingia and after a peg walked round the garden with Johnny. To my mind it is not looking well. We got home about 11 p.m.

Mon. Apr. 8th

Tested the boiler with the engine pump to 85 lbs. and the tubes hardly leaked at all but the cracks in the firebox shell are showing up badly. Ran off the water and commenced plugging.

Wed. Apr. 10th

Raised the pressure to 150 lbs. and there was scarcely a leak. At 9 a.m. Johnny came over and made a thorough inspection and caulked the remaining leaks; then he set the valves and while steam was being got up we went to breakfast. Dr Smith had come over and we were all going to polo but rain prevented it. After breakfast we ran the engine and got the governors to work after a fashion.

Thurs. Apr. 11th

First day's plucking. Started tipping the gardens and got twenty-one maunds.

Fri. Apr. 12th

Started manufacture at last, which marks another step in the season. All running fairly well. Dunlop and I went to see the canal and got water through it for the first time; it has quite a satisfactory fall.

Sun. Apr. 14th

A cold drizzling morning so I didn't go out. Dunlop returned at 7.30 p.m. with Halliday, the DC and Chrystall. I dined with them and played bridge.

Fri. Apr. 19th

Dunlop and I went over to Mijica tennis. At Mijica a big crowd turned up, including the Macs and Miss Gibb, Walter and Frank Edwards, Dr and Mrs Smith. We had a nice afternoon and when it got dark adjourned to the drawing room for music. Miss Gibb sang, but her voice was not as good as I have been led to expect and I don't care for Scotch ballads. Crutwell gave quite a good exhibition in 'Songs of Araby' and rather surprised me. Miss Gibb is leaving for home directly so we bade her a last goodbye. Reached home at 9.40 p.m. a ghastly hour for dinner.

Mon. Apr. 22nd

Plucking again and three acres of jute planted.

Wed. May 1st

A fine morning for a change and more like summer. Dr Smith came over, bringing Mears, and I gave them tiffin at the burra bungalow. Dunlop and Johnny arrived at 2 p.m. and we all went to polo. It was very hot and only two chukkas were played.

Sun. May 5th

Dunlop had a touch of fever so neither of us left the garden. I, however, went on a fruitless tiger shoot, which was brought about as follows. A lot of mundas were sheltering in the jungle and while Chema was walking down a narrow path he met the tiger coming the other way. Neither had time to get out of the way so the tiger knocked him down and mauled him slightly. He told me that as it lunged past him he caught it by the ears and if he had an axe he would have killed it. When I heard about it I went down with Budhoo sirdar and we had the jungle beaten but the tiger broke back and only the Nepali saw it. We then had another beat but with no result, so I gave it up.

Fri. May 10th

The jubilee of the outbreak of the Mutiny and things seem to be tending that way again as riots are everywhere occurring, and the papers only preach sedition.

Mon. May 13th

Went round with Dunlop. Leaf about the same as last week: twenty eight maunds.

Fri. May 17th

Dunlop and I went to Mijica tennis. Pym of WM was staying there. He is doing a round of tea tasting. Butcher, Edwards and Johnny were there and also Mrs Macrae. Play was only medium and the afternoon was oppressive, being our hottest up to date – 91° in my verandah. I had my punkahs put up in consequence.

Sat. May 25th

At 3 p.m. left for Partabghur polo ground on my way to spend the night with John Duguid. Drove Bishop to Burjaon and found out that the crossroad was not as bad as I had expected. Drove Lettice to the ground and after passing Ghope Sadarn noticed that Barker had abandoned a lot of tea, which at one time was bad with mosquito blight. On arrival at the CB discovered John waiting for me and we had a yarn on the various ways of increasing our pay by jute cultivation, etc. He has been doing a bit for several years and said that potatoes also paid a big profit out here. We turned in about 10.30 p.m. and I told him of my intention to get married, as I think it is about time to let people know.

Sun. May 26th

Had a look at the tea house and duly admired the new boiler (a Britannia) and engine. They have got a lot of machinery now, including seven rollers, three paragons and one Victoria, one Venetian, one downdraught and one updraught. Then we had a turn in the garden where the coolies were at work, as Wednesday is their leave day. At 3 p.m. left for Borpukhri and found Mrs Mac. in the verandah and told her of my engagement. She was, of course, very interested and inquisitive. Left at 6.30 p.m.

148

and called on the doctor and found him just back. Mrs Smith is away so to keep him company I stayed to dinner and didn't get back till midnight, as Mongra who had been sent to Kolapani with the Bishop, got funky and took him on to Gingia. I therefore had to drive Lettice all the way.

Mon. May 22nd

Greenfly bad and very little leaf.

Wed. May 29th

A very heavy storm of rain in the afternoon and it seems to me that the rains have broken but the weather is still cool and this year we have not had the usual spell of intense heat between the *chota* and *burra bearsat*, which generally occurs during the last week of May and first week of June. Swinley sent carts and borrowed some coal.

Mon. June 3rd

Went up early and had a talk with Dunlop; then, after *chota*, we went round Monabari and he was disappointed not to see more leaf. At 2.30 p.m. Daloo reported that a tiger had killed a cow of his in the Daria Khet but had not touched it so I sent him out to get a pit dug and also the *mistries* to make an entanglement of barbed wire. Having summoned the Nepali, I set off with rifles, etc., and on reaching the place discovered the cow lying in grass about three feet high. All the *busti* wallahs were there and very soon a pit was dug and ready to put wire round it, whereupon Buddhoo sirdar and I took up our places. After waiting about an hour we heard the tiger growling but it didn't come near. Then the Nepali arrived and I made him sit slightly to the rear of us. We continued thus until dark, when I decided to abandon the attempt. Dunlop had sent the Bishop for me and I rode home.

Tues. June 4th

Very wet and towards evening the Bargang came down in flood. The new DSP Gordon came to stay the night. Our first days packing of the year.

Wed. June 5th

Borrowed Gordon's *hathi* to cross the river and see the drain. We started at 6.30 a.m. in a slight drizzle and entered the river opposite the brick kiln. The water was tearing down and this was my first experience of crossing in a flood. It put me in a blue funk for when we got three parts over the *hathi* stopped and seemed unable to go forward, while the mahout said it would be worse to turn back. However, with sundry lungings we reached the bank and soon arrived at the drain, which was running very well. We went up to the inlet and I settled to recourse where the old channel meets the present one opposite the Daria *busti*, as the channel was broader; but when we got halfway it suddenly got very deep and the *hathi* sank down to swim. The current took us into rough water and several times waves came over the pad and soaked me. We soon got out, however, and I was not nearly so nervous as on the outward journey. We reached the bungalow at 9 a.m. in a perfect downpour. Gordon left before tiffin. It was too wet to turn the people out and altogether over four inches of rain fell.

Thurs. June 6th

Still raining so I ordered a gang of Mundas to work on the drain and accompanied them myself. We crossed in

150

the boat and as they are not good at working it we got broadside in the rough part and endured a decided pitching and tossing. It was slow work walking through the jungle but eventually we reached the inlet and set to work widening the mouth by hoeing the sides down and letting the rush of water take the silt away. During the afternoon I went to Kationabari. We have had seven inches of rain in the last three days. We let the women in early as it was bitterly cold and still drizzling.

Fri. June 7th

Dunlop had a fire at midday, which shows this is a record year for low temperatures. Dunlop went to Borbheel for the night.

Thurs. June 13th

Dunlop arrived full of the Catto case. Catto will lose his case and it is costing him Rs. 500 per day in solicitor's fees. Up to date he had spent about Rs. 5000. The garden will pay for him, of course, but it is a big sum to lose over stealing coolies.

Fri. June 14th

Agnes wrote to say she has arranged to come out on the *Sardinia*. I told Dunlop and, of course, he wasn't very keen but can't help himself. The Imperial Tea Company report is out and he has made nearly £7.700 profit and drawn nearly £400 commission.

151

Sun. June 16th

Dunlop drove me to breakfast at Borpukhri. At Mijica we found Swinley and Crutwell in the office and had a talk with them. At Borpukhri a regular crowd had assembled: Mrs Cooper, Miss Robinson and her brother, Edwards and Butcher. Noble has just announced that he is engaged to be married to a Miss Pengrie, so this subject afforded some discussion.

Mon. June 24th

The leaf has come away with a rush and today was very hot. I worked on the firebox, which had started leaking again, and when I came into the open air at noon it seemed cool and I ordered the leaf into the top *chung*. But when I found my bungalow thermometer standing at 95° I saw I had made a mistake and so had it taken down again. During the afternoon got all leaks stopped. Leaf 140 maunds.

Wed. June 26th

My whole time is now taken up in the tea house. Today's total: 209 maunds.

Fri. June 28th

I shall be glad when this heavy flush is over as the last few days I have not been in bed before 11 a.m. and have started firing at 4 a.m. We have made an office in the tea house, which fills a great want.

Sun. June 30th

Dunlop and I intended to go to Majulighur, but it rained heavily, so we went to Gingia instead. Johnny was plucking and getting plenty of leaf.

Sat. July 6th

Still plucking, but only got fifty one maunds. Dunlop went to Gingia for the night. We have had over twelve inches of rain this week.

Friday was the anniversary of the first day I met Agnes. How well I remember the consultation with Mary as to getting a fourth for tennis and finally taking notes for the Pendleburys and Boulter Cooks. Riviera was the nearest house and Agnes took the note. I felt strangely attracted and when Rita came to play I was terribly disappointed till Mary made up for it by arranging a picnic.

Sun. July 7th

Only fifty maunds to manufacture. During the afternoon we sampled teas, Gingia, Majulighur and Monabari. The strongest were Gingia.

Sat. July 13th

Short leaf again. It has drizzled nearly every day this week and the carts have only twice been able to cross the Bargang. The river is fast cutting No. 4 away and if the bund goes we shall lose 5 and 24 in addition to all the dhan land.

Sun. July 14th

Showery day. Johnny came to breakfast. The tea house closed at 7.30 p.m.

Mon. July 15th

Drove to Burjaon to inspect the road against the arrival of our boiler.

Wed. July 17th

Dunlop went to polo but returned early as he had received a wire saying that the boiler would probably land tomorrow. We made all preparations for going down, including gharries, *mistries* and our private messing in the floating dak bungalow.

Thurs. July 18th

Sent off Arkitia with twenty coolies, followed by nine carts and at 11 a.m. we left ourselves. We passed a lot of jute beyond Pani Boul where the Bishnauth plain drops down to the *dhan kets*, which evidently belong to the Mymensingh people. I had only once before landed at Bishnauth and then it was dark, so Dunlop pointed out the different roads: one to the south, another through the Doon *busti* and the third a straight road running to the top of the hill which drops rather steeply to the water. We went to the Kyah's shop and put up the buggy; then through the Doon *busti* road to the ghat, where we found the flat and our boiler in the water beside it. We went on board and made friends with the mate of the *Krishnagar*, while his men rolled the boiler onto the sand. Then we repaired to the floating dak bungalow, where Feringhi Minchand and Rengit had

154

turned up and had a feed at 6 p.m. The evening was cool and we required no punkahs.

Fri. July 19th

Started early, jacking up the smoke box end, and by noon had got the wheels at the firebox and rolled in underneath. Then, in the afternoon, we dug trenches for the wheels and, after getting them in position, jacked them up, put planks underneath and set the elephant to pull, which it did very ineffectually. So we sank an anchor in the sand and by this means got the boiler onto the spread of planks laid down for it. We left it thus for the night and Dunlop sent for the Majulighur *hathi*.

Sat. July 20th

Started early with the block and tackle lent us by the mate of the *Krishnagar* and pulled the boiler over the sand as far as the mud and then had chota. Afterwards we planked the mud. At 11 a.m. the *hathi* arrived and as the mahout was a smart chap we got across by tiffin time. After Dunlop left I commenced the ascent with one *hathi* pulling and one pushing behind, but they didn't work in unison, so the wheels dug holes for themselves and the continued jerking canted the boiler over. By nightfall we had not straightened it.

Sun. July 21st

Straightened boiler with the jack and decided to run no more risks but plank the whole length of the hill and pull up with the anchor and blocks, using the *hathi* as well as the coolies on the rope. This way suucceeded surely but slowly and at night we had reached the first level near the

155

bridge. Weather has been fair but the river rose and I now have to wade to the bungalow. On Friday night the rain came through the roof and I woke to find my bed very damp.

Mon. July 22nd
Put iron beams across the bridge and then put planks over them so nothing broke. At noon we were half way up the next slope. When I returned to the bungalow I found, to my disgust, two PWD babus there, one of whom proved to be Mullick, the engineer. While I was washing Dunlop arrived, to my surprise, and said he had brought twenty coolies. He talked to Mullick, who allowed us to take the anchor as far as the garden and told us to go dead slow over the bridges, as our weight of 150 maunds' boiler and fifty maunds' wheels was about the limit. After tiffin we got all forty coolies on the ropes and pulled up two more lengths, after which Dunlop left. At 6 p.m. I reached the top, when the *hathi* was hitched, and with the tackle this time behind we let the boiler down the slope, loosed the rope and the *hathi*, unaided, took it a few hundred yards along the road.

Tues. July 23rd

Paid off the Majulihur *hathi* according to agreement. Left at 8 a.m. and progress was very slow as the Behali *hathi* kept stopping every twenty yards and wouldn't start without all the coolies pulling on two side ropes. Another cause of worry is that four coolies have got bad fever and there is no place to put them. At midday we had progressed about a mile and then got stuck in a mud hole, to get out of which we planted the anchor. This was opposite the Muslim jute colony, which they had begun to reap. They were very civil and showed me the fibre. I tried an experiment of hitching two buffalo carts to the ends of the

156

side ropes and found it answered well. We now got on much better and by sunset had pulled half way up the slope at Pani Borel. I wanted to reach the top, but everyone was dead beat, so I returned to Bishnauth for the night.

Wed. July 24th

A pouring wet morning and the road was very slippery. The boiler skidded across the road till it was stopped by the high bank. By noon we reached the top of the hill and pulled a short distance along the road. My fever patients were now better and the sun came out strong. Drove back to breakfast and arranged to move my camp to Butigigaon; also paid all bills and distributed baxis.

The river was now quite high and I had to use a boat. I wasn't sorry to see the last of the ghat, though things might have been worse. But on the whole it was a cool place and there were two beautiful moonlit nights, which showed the river off to perfection.

Reached Pani Borel at 3 p.m. and found another sixty coolies so things ought to move now. Dispensed with the buffalo and started to pull using the long ropes from the steamer. The heat was now excessive and every half mile or so I had to give the coolies a rest. To add to their discomfort, there was no water as we were now on the Bishnauth plain. About 5.30 p.m. we drew near the Pukri and to my astonishment, I met Dunlop, who had read a chit of mine incorrectly and understood I was to be at the Burigaon bridge. However, he was able to see the difficulty of my task and after a little conversation he went home. I had now to tackle the bad mud hole at Charali hat, but the coolies got the boiler on a run and had no trouble in getting through and turning on to the government road, where at 7.30 p.m. we stopped for the night and I drove to Burigaon rest house, hardly able to speak for thirst. I had to spend the night without a punkah but luckily it was not over hot.

157

Thurs. July 25th

Rose at 5.15 a.m. and reached the boiler at 6 a.m., where the coolies were waiting. We got on well for a short distance, when it came on to pour. However, we got to Burigaon by 8 a.m. and by this time everyone, including myself, was shivering with cold. I had breakfast but the coolies could get no firewood and took their food uncooked. At 9 a.m. it cleared up and we restarted, as another fifty coolies has arrived.

In order to cross the big bridge as slowly as possible I planted the anchor and by using the tackle we kept her moving. I allowed no coolies on the bridge, except about half a dozen to guide the wheels. When all the weight was on there were several ominous cracks and the ironwork quivered, but we got safely over and I drew a breath of relief. After this I sent Girlish and the *mistries* to put a support in the Gingia bridge and returned myself to pack up at the rest house. On rejoining the coolies I found them half a mile beyond the tank and we got on well till we crossed a culvert, where the hind wheels broke through and sank, till the boiler firebox rested on the road; but by digging trenches and getting the elephant to push behind we heaved her out. We crossed the bridge at Gingia, the only mishap being the sinking of a hind wheel when touching the first crossbeam. But the elephant got us out of this and the rest was plain sailing. When we passed the direct jungle road to the garden I knocked off all and went home. Dunlop was delighted with the result and I went to the burra bungalow for some much needed refreshment. I also dined with him and cracked a bottle of the 'Boy'. Now that the strain is over I feel as if I could sleep for a month.

Fri. July 26th

The coolies ate their rice and went out at 7.30 a.m. so I

had my chota in comfort; then rode the Bishop down the old road, which was not as bad as I expected. We started pulling at 8 a.m. and, there being a strong sun, the heat in the low part of the road was simply sweltering. But when we reached the garden the paniwallahs arrived and I rested the coolies. Dunlop joined us here and we pulled on to the tea garden. Then we rested the coolies, while Dunlop and I went to the bungalow for a drink, it being nearly midday and over 90° in his verandah. We returned later to find the coolies had pulled on and the boiler was safely in its place at the tea house. So we gave everyone leave and went in to breakfast, very thankful that the job was at last over.

Sat. July 27th

Returned the Behali *hathi*, straightened the boiler and jacked it up, moved the wheels and let it down on the planks. Cleaned the rope blocks and dispatched them to the ghat. Only fifty maunds of leaf.

Sun. July 28th

Helped Dunlop make preparations for his big spree.

Tues. July 30th

Completed all arrangements and at 3.30 p.m. went to the burra bungalow. Guests arrived shortly after, Butcher on a *hathi*. Tennis went well till dark, then we all changed into soft shirts and white dinner jackets. A total of seventeen sat down to dinner. There were three dozen fizz and one dozen brandy to be mopped up. A long punkah had been rigged on the verandah and we had two tables there. Supper was started at 1 a.m. and by 2 a.m. everyone was off to bed.

Wed. July 31st

Rose at 5 a.m., and as Barker was awake, we went down to the tea house, afterwards returning for tea, when the other men tumbled out. *Chota* was served at the burra bungalow about 8 a.m., and terribly mixed drinks were indulged in, such as stout and champagne, Munich beer and Allsops etc. Tennis followed. Shortly after noon a start was made for the polo ground. At Kolapani the rest of the district were assembled. Swinley had arranged a splendid tiffin and a placard of entertainment was announced, one item being that Cragoe would sit for his portrait, copies of which would be sold and the proceeds devoted to buying new riding breeches for Johnny Dey. After tiffin we had bridge, followed by polo. No one was late leaving, as it clouded over and we reached Monabari in the rain. The whole entertainment has been an unqualified success and on the biggest scale I have known; the cost will not be under Rs. 400.

Thurs. Aug. 1st

At work again and in no particular mood for it. Our return is just a few maunds over last year. Home mail and Mrs P. says she won't let Agnes travel from Bombay alone, so I don't know yet when she will arrive.

Sun. Aug. 4th

Dunlop was taken ill with violent diarrhoea and vomiting and nearly collapsed. Johnny came over in the evening and took him back for the night.

Mon. Aug. 5th

Spent the day laying off the foundations for the new boiler

and riveting up the old. Dunlop returned at night.

Tues. Aug. 6th

Dunlop feeling better. The doctor came and prescribed.

Sat. Aug. 10th

Home mail arrived but Agnes has not yet definitely settled on her passage. Paid the Kationabari coolies.

Fri. Aug. 16th

Still hot and dry. Dunlop and I visited the nursery where we found some bad work. Dunlop has a touch of fever.

Sat. Aug. 17th

Half an inch of rain fell, which ought to do a bit of good, but leaf fell to sixty three maunds. Godwin arrived at 3 p.m., looking very fit. He has not been here for at least a year and a half. Johnny came over so we dined at the burra bungalow and mopped up some bottles of the 'Boy'.

Mon. Aug. 19th

Dunlop seedy. He had a fall and semi-fainting fit just before going to bed, which left him very weak.

Tues. Aug. 20th

Dr Smith examined Dunlop and put him under treatment.

161

Thurs. Aug. 22nd

Leaf good, 191 maunds. Godwin went to Bishnauth Ghat. Home mail. Agnes says she has booked her passage on the *Sardinia*, which is due nine weeks today.

Wed. Aug. 28th

Rained heavily at 4 p.m. and Dunlop did not go to polo. Leaf growing fast.

Fri. Aug. 30th

Dunlop seedy again. He now talks of possibly going home next year.

Mon. Sept. 2nd

Connected the new boiler, got steam up and ran the engine. Everything appears to be in order. Leaf 180 maunds.

Wed. Sept. 11th

Went to polo for the first time for six weeks. There were only five players and no ladies. On the whole it was not worth going down for.

Wed. Sept. 25th

Dunlop went to polo. I have developed a wretched cold. The weather is vile, dull with perpetual showers; absolutely unlike September.

Tues. Oct. 1st

Leaf shows signs of closing up and the bushes have suddenly developed a cold weather appearance. This morning the temperature fell to 70° and my early morning ride to Kationabari was very enjoyable.

Thurs. Oct. 3rd

Wrote to a Calcutta padre about the wedding and also to book a room at a boarding house for my four days' quarantine.

Sat. Oct. 5th

Sent the buggy for Dunlop but his steamer was late and he didn't arrive till 7 p.m. He is feeling better for his short stay at Tarajuli. The secretaries have written to Dunlop regarding extensions and we are to put out 200 acres. This will mean a lot of work next cold weather and rains.

Thurs. Oct. 10th

Biked to Borpukhri for breakfast and at Mijica found Swinley, Crutwell and Burton in the tea house. Afterwards called on Mrs Mac. and the baby. Stopped for ten minutes or so and arrived at Borpukhri as the women were weighing their leaf. I felt a bit fagged and was glad of a bath and change, after which Mrs Mac. gave me a list of things to buy, and then we had breakfast. They are giving us some dessert things, which is very good of them. I heard that Dunlop had discussed arrangements and announced that we should dine with him on our arrival, but Mrs. Mac. is to put a stopper on that, thank goodness. They are to come over the following day, which will be very nice as

Agnes will be pleased to see another woman. I left at 3.30 p.m. and reached home 5 p.m., feeling absolutely done, but managed to pay part of the new lines.

Fri. Oct. 11th

Got a dash of fever at noon but the doctor happened to be over, which was lucky, as after breakfast my temperature rose to 102°. He says it is just funk and I think it quite likely.

Sat. Oct. 12th

Feeling much better, though weak. Visited land at the back of the burra bungalow, where we propose putting out our 200 acres. Paid Kationabari in the afternoon.

Mon. Oct. 14th

Johnny sent me a cheque for Rs. 120 in the names of Swinley, Butcher, Crutwell, Bradden Robinson and himself.

Tues. Oct. 15th

Started packing. During the afternoon Podo came and told me a jungly buffalo was eating his *dhan* in the *pathar*, so I went out but it had gone, which was a pity as the position was excellent for a shot. Dunlop is giving me Rs. 100 as a wedding present.

Wed. Oct. 16th

Dunlop and Bradden shot over Kationabari, but as I expected, only got two brace. Finished packing and made arrangements for dinner and the night at Behali rest house. At 4 p.m. Dunlop and Bradden went to polo and I drove the Bishop to the ghat. I crossed in the dugout but the buggy was led through. I put up in the rest house and had the pleasure of witnessing the last of the Durga *puja*. Kodo provided me with a fairly eatable meal and I brought the dhobi as my servant on account of his knowledge about Calcutta.

Thurs. Oct. 17th

At 2 a.m. the boat arrived and I got a cabin to myself. Had *chota* at Bishnauth. At Tezpur Morgan came on board for a few minutes and Whyte, the engineer. Travelled to Gauhati, which we did not reach till after dinner, as tea was loaded at Singri and Mangaldai.

Fri. Oct. 18th

Left Gauhati at 3 a.m. A lot of men joined us last night and one lady. Two were in the forest department and one, by name D.L. Stewart, in jute. At Dhubri most of us landed and I got a carriage with Stewart, thus being able to hear a lot about his business, which is just being started. He virtually offered me a billet on Rs. 400 and I shall think about it. He also offered to buy my car. We soon passed Gitalda, which is the junction for Buxa, and at Lamoni stopped for half an hour to get dinner. The refreshment room is well run by the Parsi firm of Sorabji.

Sat. Oct. 19th

Dressed before daylight and just when the first flush of dawn appeared we reached Sara Ghat and went down to the ferry, which is only a hundred yards from the platform.

The boat was a good one and *chota hazri* was laid on deck, so we had an ek dum meal. The Darjeeling mail arrived with about forty passengers and we got under way. The Ganges didn't appear very broad but the light might have been deceptive. At Damukhdia we managed to cram ourselves and our luggage into a carriage and about twenty minutes later we departed. There was nothing remarkable about the scenery, but in the carriages an electric bell is fixed so that by switching it on at night it rings if anyone opens the door. This has been put on since the robberies.

We reached Sealdah about 10 a.m. and after getting a gharry I drove to my quarantine place, 16 Mangoe Lane, which is a seond rate boarding house, but as I have a fan in my bedroom I can stand it for a few days. I changed and drove to the office where I found Edwards; then saw Bowrey and Cassie about the boiler and we had tiffin in the office.

Sun. Oct. 20th

Edwards turned up at 7 a.m. and took me for a drive in his tumtum. We went to the zoo and this time I found some fine king cobras. There was also a baby elephant, a small mithun and the rhino which Edwards had not seen before. Got back about 10 a.m. with an invite to tea and dinner.

I spent the morning writing and after tiffin loafing till 4 p.m., when I took a gharry for No. 1 Camac Street. Punctually at 4.30 p.m. Edwards was waiting for me and conveyed me to his rooms (it is a boarding house) and introduced me to his wife who, not actually pretty, is very attractive. They were engaged six years, three before he came out and then three of absolute separation. I mentally salaamed to their constancy. We had a splendid tea and at 5.30 p.m. Edwards drove to the Old Mission Church. The evening was beautifully cool and we let the pony walk down the Red Road, the fashionable thing to do. I had never

166

been in an Indian church before and this was better than I had expected. The congregation was very mixed, likewise the choir, which was only passable. The Rev. H. Clark is an ascetic-looking person and preached a good sermon. I took stock of him and also the building as the words spoken by him, myself and another in this place a few days hence will make or mar my life and it will be my fault if they mar it. We interviewed him afterwards, finally returning to dinner, which was served below and was absolutely first class.

Mon. Oct. 21st

Did a good morning's work at Harnack's, etc. Called at the office, were I found a letter from Agnes written from Aden. She said the skipper expected to arrive on the 22nd. I hurried to the P & O office and there, to my disappointment, was told that she would not be up till Friday at 5 p.m. This will cut our Darjeeling trip short unless we can be married before 7 p.m. After tea I went to Hamilton's and bought two rings; one engagement and the other the wedding ring. I engaged rooms at the Grand and it seemed very strange to enter Mr and Mrs Hetherington.

Tues. Oct. 22nd

Spent most of the morning at Whiteways and Osler's. My bills are getting colossal. In the afternoon went to the Radha bazar and found the natives much more civil than I had expected.

Wed. Oct. 23rd

Found a letter from Agnes, saying that she won't accept the Edwards' invitation but is going to the Grand with the

Hon. Mrs Napier and also that she won't be married on Friday, as it will be too late. This absolutely upset all my plans and I had to cancel our suite of rooms at the Grand and book her a single room. Then I went to the Great Eastern and booked a suite but had to pay Rs. 5 per day as a retainer. I also engaged for myself a room at the Grand. Went to a furniture shop in Bow Bazar and got a dressing table, two washing stands, an almirah, etc., for Rs. 140. Paid my bill, which was only Rs. 20, and left for the Grand. Called on the Edwards and we discussed things. The best arrangement seems for us to be married on Saturday, about 5 p.m. and start for Darjeeling on Sunday afternoon.

Thurs. Oct. 24th

Received Dunlop's present of Rs. 100. Chota is very late here, which curtails the morning. When I was having it a man came up and spoke. He proved to be Wood of Doom Dooma, with whom I had some business over coolies in 1905. He was very friendly and thanked me again for the way I had helped him, though it was very little. We sat together and he told me he had come down about his eye. After *chota* did a big round and made an appointment with Milton to try a pony during the afternoon. Also ordered wedding and visiting cards. At 3 p.m. arrived at Milton's and he showed me a grey mare over thirteen hands, which he put in a dogcart and drove me out. I also drove myself and the pony has a fine mouth and is also quite quiet. At the finish I decided to call again. While arranging a room for Agnes near Mrs Napier I was accosted by Napier himself, who asked me to sit at the same table as himself, which I did, and we arranged that he will go down on the Customs launch, as he has a pass, while I have gharries ready at Chandpal Ghat. He will get the luggage passed and separated out so there should be no trouble.

Wrote a letter to Agnes to send in charge of Cook's man. Got a wire from the old man, offering me the Bishop at Rs. 600, which I don't think I will accept. From tiffin till 5.30 p.m. I loafed, then I had tea, changed and took the bearer in a gharry to Chandpal Ghat, where I arrived about 4.20 p.m. and was told the *Sardinia* would not be up before 5 p.m. As I had brought my new Prismatic binoculars I drove down to the riverside to look out for her but could see nothing. I then returned to the hotel to leave my topee for a straw and got to Chandpal about 5.30 p.m. Here I found Shirley Hodson and a crowd of others, while the *Sardinia* was just making her appearance.

I used my glases and also lent them to him, but neither of us could discover Agnes. The boat was very quickly warped in and just when the gangway was erected I saw Agnes and by a fair amount of pushing I succeeded in being the second man on board.

We entered by the porthole and I found her on the hurricane deck. We went to her cabin, collected all the light luggage and returned on deck, where Captain Talbot came up and spoke; also the third officer, who is a brother of Vincent Ward. Shirley Hodson, in a thoughtful way, had absented himself and so had the Napiers, so we went on shore and after some difficulty picked up my ticca gharry and drove to the Grand.

I told her on the way my arrangements for the wedding, but she had got a bridesmaid, been lent a brougham and issued several invitations with the idea of being married in the cathedral. So I decided to go at once to Edwards and find out if we could change the venue.

On reaching the hotel I took her to her room and departed straightaway for No. 1 Camac Street, where Edwards and I arrived simultaneously. His memsahib was out, so we had a peg while I explained matters, and then he drove me to the cathedral. It was shut up but we found the verger, who took us to No.1 Theatre Road, where one of the padres, the Rev. Dr Cogan, lives. He was out but I made an appointment for 9 p.m. Then Edwards drove me to the Grand. Having dressed for dinner, I met Agnes on the landing and was introduced to Mrs Napier and a friend of theirs. We all dined together but at dessert time I had to leave.

Dr Cogan was at home and gave me a cordial greeting. He is rather the type of Irish priest as drawn by Lever and we got on first rate. He explained that all I had to do was to get my licence transferred by Mr Clark, so our business was quickly concluded and we fixed the wedding for 5.30 p.m. Reaching the hotel, I found Agnes on the balcony with the Napiers, but as it was after 10 p.m. they all turned in and I was left disconsolate. My last bachelor day was over.

8

MARRIAGE IN CALCUTTA

Sat. Oct. 26th

The day dawned brightly as a good omen and at 7.15 a.m. I was off in a tram to visit the Rev. Clark at 11 Mission Row. He told me Agnes would have to be present when we applied for the licence so I returned for her but she was a long time dressing, so we had to *chota* first and then went down. At his house we met a *Sardinia* girl who was about to be married, but he gave us our licences and we got back about 10 a.m. After this I went to the New Market to fetch rackets I left to be repaired; they were well done and cost Rs.4/8. I also ordered bouquets for the bride and bridesmaid. Called on Shirley Hodson and arranged with him to be my best man; then to the Customs, where I could only find a small amount of luggage, though I searched through the warehouses till perspiration and bad temper exuded from every pore. Called at the office, where I was given a silver box lined with cedar to keep cigarettes in.

The rest of the party were at tiffin when I turned up and I didn't have time for much, as at 2 p.m. I drove again to the Customs and finally took a dinghy to the *Sardinia*, but could not get delivery of the silver box. Reached the hotel at 4 p.m., where Shirley was waiting. My frock coat, etc., had arrived from Edwards. I changed and everything fitted well, barring the boots, which had to be stuffed with paper. At about 4.50 p.m. we drove off in great style and soon drew up at the cathedral. I then paid the verger the necessary fees and he took us up the aisle to the front seat on the decani side and explained what we had to do. After a few minutes Agnes arrived, escorted by Miss Egremont, her bridesmaid, and Mrs Napier, so Shirley and I moved to the front and took up our position. I was not a bit nervous, contrary to my expectations, and we got through the ceremony all right, finishing by signing our names quite legibly. The Captain and Ward, Mr and Mrs Edwards, Mr and Mrs Matthews and Bayley a brother-in-law of Miss Egremont, were present and gave us a good send off in Colonel Sanders' brougham.

We drove round by the Red Road to the hotel, where we packed up and set off for Pelitis. There I left Agnes and took the luggage to the Great Eastern. Returning to Pelitis, we had not long to wait as first Shirley, then the Edwards, and finally Bowrey arrived. The dinner was fairly good and they left us about 9.15 p.m.. Our bedroom looks out over the Old Court House Street and we sat on the balcony for a little while before turning in.

Sun. Oct. 27th

Rose about 7.30 a.m. and a very hot morning it was; also, to our disgust, the punkah had not been erected in the sitting room, so we moved the beds and had *chota* in the bedroom. About 10.30 a.m. I called at the office for letters and then drove to Sealdah to book sleeping berths in the Darjeeling mail. On my return the punkah was erected so we had breakfast there and then wrote letters till 3.30 p.m., when we had tea and immediately after left for Sealdah. Shirley met us there and I had a little trouble over the luggage, but all was fixed up before we left at 5 p.m.. The journey to Damukia was not as interesting as I expected, for Agnes had been up to Kandy and the scenery in the plains is very similar; but the crossing to Sara was weird and the electric light very pretty. At Sara I found a mistake had been made in reserving our berths, as they were both

172

labelled Mr, and another man was in the same compartment, so Agnes went to the ladies carriage. We left about 9.30 p.m. and I turned in immediately.

Mon. Oct. 28th

Woke at 6 a.m. and found we were running on a dead level, straight road towards the hills. There was one fine snow peak showing and the sun, which was just rising, threw a pink glow on it. We reached Siliguri about 7 a.m. and took our luggage to the DH train, which was a quaint affair and more like the Tezpur railway or a steam tramway than a railway. We had *chota* in the refreshment room, run by Sorabji, who are very smart caterers in every way. I had put on my warm clothes at Siliguri and found them uncomfortable till we started.

The journey was most enjoyable. First we got into forest where there was no view but masses of creepers and flowering shrubs. Then we came out on the hillside overlooking a large valley and we went round a couple of loops where the track winds overhead. Suddenly the train stopped and commenced backing and we zigzagged several times till we had risen several hundred feet. After this there was not much change till just before Kurseong, when an uninterrupted view of the plains burst upon our eyes. We were now over 4000 feet and it was a most marvellous sight, resembling a great map, the far distance lost in a misty haze.

At Kurseong we had tiffin and on leaving there we turned into valleys again. In places there must have been a drop of a thousand feet sheer down from the track. We passed several gardens. How the women pluck is a mystery to me, unless they hold cn with one hand to the bush. About 12.30 p.m. we reached Ghoom, the summit of the line, and shortly after ran down into Darjeeling. Here again the luggage gave trouble, as it was set upon by hill people, mostly women who wanted to carry it in different directions, but the bearer came to the rescue, so Agnes and I set off in a rickshaw to the Rockville Hotel. The ascent through the town was very steep and on our way we passed the Woodlands, then a sort of square with a bandstand and fountain. Finally, after the steepest bit of all, we reached the hotel. We found it very full and we could only get one room with a corner curtained off for a dressing room, instead of a suite as I had ordered. There was a fire, however, which made things cheerful, though it was not as cold as I had expected. We had a second lunch, then unpacked. At 3.30 p.m. we had tea and afterwards walked to the square and back, but it was cloudy. Dinner at 8 p.m., after which we sat by the fire, which was very enjoyable.

Tues. Oct. 29th

I got up about 6 a.m., and found there was a magnificent view of the snows with Kinchinjunga in the centre. The air was chilly, so I went for a walk till Agnes was dressed, after which I brought her to see the view, and then we had *chota*. Afterwards we walked down to the bazar and bought a ring with an opal heart set in diamonds. Agnes returned in a dandy, while I walked, and on reaching the hotel fixed myself up and returned to be photographed, which I found to be rather an ordeal. We had an early tea and at 4 p.m. took rickshaws, going down past the Woodlands and then up to Jelapahar. It was a very steep ascent and when we got a good way up the view was grand. Then a great wall of mist rose from the valley and the sun shining through it produced an extraordinary effect. At one point our rickshaws had to make way for a fat lady in a dandy, who turned out to be the maharani of Cooch Behar, whose house we later passed. After this we soon reached the top and turned into a parade ground where some tommies were kicking a football about; then we came round a corner, and suddenly the snows appeared free from mist and with a soft rosy glow on them. They were only visible

174

for a few minutes and then obscured by cloud, but they were worth coming any distance to see.

Wed. Oct. 30th

A cloudy day. Went to the bazar again. At 12.45 p.m. we had tiffin and dispatched our marl and soon followed ourselves. I had engaged seats both on the DH and at Siliguri, but we did not get nearly such nice ones as on the way up and the train was crowded. We left at 2 p.m. and as the afternoon was misty the view was not nearly so fine. We had tea at Kurseong and reached Siliguri at 7.35 p.m.. Here again we had trouble in finding the luggage but at last we put it into our sleeping carriage and went to get some dinner. The train was an hour late in starting and throughout the night there were frequent delays. However, we were in a through carriage and so didn't have to change at Parbatipur.

Thurs. Oct. 31st

Woke to find us not far from Lamoni and when we reached it got some eggs and bacon, which were much appreciated. Arrived Dhubri at 10 a.m., and to our disgust found our marl not there, as the station said it had been taken out at Parbatipur. The steamer was waiting so we boarded her and she started jat pat. Then I found it was the special Gauhati boat and the dispatch boat was ahead. I now regretted I had not waited at Dhubri, and when we came upon the down boat stuck on the sand and the dispatch boat pulling her off I got the *serang* to put us on board the dispatch boat. There, to my surprise, we discovered Sleepy and Cooper. When the down boat had been pulled off she came alongside and we went on board. We got back to Dhubri at 8 p.m. and at 11.30 the bearer came and said the luggage had arrived.

175

Fri. Nov. 1st

At 10 a.m. the *Afridi* arrived and we embarked in earnest.
Two people came in by the mail; I think they were
Edwards, the Darjeeling solicitor, and his wife. We reached
Kholabanda at night.

Sat. Nov. 2nd

There was a dense fog at daybreak and we didn't get up
anchor till about 9 a.m. We reached Gauhati at 5 p.m. and
the view was lovely – much better than I have ever seen it
before. We went on shore and posted letters home.

Sun. Nov. 3rd

Boat ran all night and we arrived at Tezpur up to time.
A red-faced man got on board there, who turned out to
be Catto. He told me Bruce had made a record season at
Hattigor and Willie Briscoe had made a loss at Dimakusi.
Mrs Ernest Holden and the Nowgong DC, Edwards, came
on at Gauhati. He is anything but like a government
official. We reached Behalimukh at 5 p.m., where Hirisai
and Lettice were waiting, and an hour later we were at the
bungalow. It looked quite respectable but there was no fire
as I had expected. We went to call on Dunlop and found
Johnny there. They told us Miss Robinson, her brother,
Butcher and Mitchell had come over the day before to
meet us and worked at the bungalow. We dined at the
burra bungalow but slept in our own.

Mon. Nov. 4th

Chota at the burra bungalow and afterwards I went round
with Dunlop. The old boiler got too bad for work on

Saturday so he connected the new one. He went to Bargang in the afternoon and we started unpacking.

Tues. Nov. 5th

Visited the nursery at Maduri Bheel, where Dunlop had decided to put out the extensions. During the afternoon I drove Agnes round to Kationabari to give the Bishop a trial, as I am thinking of buying him. Dunlop had tea with us.

Wed. Nov. 6th

There was no polo but Johnny came over and played bridge. Agnes and I were partners and only got beaten by a few points; it was our first game together. A good deal of our stuff has now arrived and we had a fire for the first time.

Fri. Nov. 8th

Dunlop went to Gingia for breakfast, on to Mijica for tennis and back to Gingia for the night. I developed fever, temperature 101°.

Sat. Nov. 9th

Still feverish. Dunlop returned to breakfast. He had fever himself yesterday.

Sun. Nov. 10th

The piano arrived and I spent most of the morning putting it in its place. Our first callers arrived in the persons of Swinley and Crutwell.

Wed. Nov. 13th

Started pay, so we were unable to go to polo.

Fri. Nov. 15th

Drove to Mijica for breakfast. Mrs Macrae and a Miss Clark from North Lakhimpur also came over. After breakfast we had a little music and then tennis, at which Robinson and his sister, Dr and Mrs Smith and Macrae turned up. It was a very pleasant meet. Dunlop returned with Johnny and Major Halliday, who has been very seedy.

Sat. Nov. 16th

Finished pay at Kationabari. The Major called on Agnes and we bridged at the burra bungalow before dinner.

Mon. Nov. 18th

Agnes and I went to Bargang tennis. We were rather late in arriving, so I didn't get a game. Swinley, Crutch, Johnny, the Major and the new Behali assistant, Bowker, whose name describes him well, were there. There was the usual singsong, in which Agnes and Crutch did most of the work. We came away at 6.30 p.m.

Wed. Nov. 20th

Dunlop went down to Tezpur with the Major and I dispatched the car to Chaston for a trial.

Sat. Nov. 23rd

Paid ticca at Kationabari and Agnes drove herself for the first time alone to meet me.

Sun. Nov. 24th

Butcher came to call about 4 p.m. and stayed till 5.30. Directly after he left Johnny turned up and we persuaded him to stay for dinner, so he was our first guest.

Mon. Nov. 25th

Met Dr Smith on my way to the nursery and asked him to stay for breakfast. He had come over to help Dunlop get through his teeth pulling as the dentist was to return with Dunlop from Tezpur, but owing to the boat being late they didn't arrive till dark. The doctor left about 3 p.m.

Tues. Nov. 26th

Dr Smith came over again and Dunlop had ten teeth drawn, cocaine being applied. The dentist, Dr Osborne, is an American and not a bad sort. He is young and cleanshaven.

Wed. Nov. 27th

Dunlop feeling a wreck. Swinley, Mrs Macrae and Mrs H. Smith came to be operated on and Mrs Smith called on Agnes.

Mon. Dec. 2nd

Usual work round and Dunlop like a bear with a sore head.

Tues. Dec. 3rd

Last day's plucking.

Wed. Dec. 4th

Drove Agnes to the polo ground for the first time and there was a big crowd. Altogether there were two women and seventeen men. After polo there was a managers' meeting to decide whether we are to have two doctors instead of one. The vote was in favour of two. Agnes and I went to Mijica for dinner and the night. Swinley took Agnes in for dinner, Lawes took Mrs Hutchinson, Mac. took Mrs Smith and Edwards Mrs Mac. After dinner we had a lot of music and Mrs Hutchinson recited.

Thurs. Dec. 5th

Got a letter from Chaston, saying he finds the crank pin broken and is rather sick at having given Rs.600 for the car. Left Mijica about 10 a.m. and on reaching the nursery I walked in and found Dunlop there. He drove me back.

Wed. Dec. 11th

Dunlop drove Agnes to Polo. Gordon, the DSP, and Master Huntley spent the night at the burra bungalow.

Fri. Dec. 13th

Met Dunlop at the nursery and in the afternoon we went to Kationabari. I finished pay. During the evening Agnes and I went to the burra bungalow to hear his new records.

Sat. Dec. 14th

Dey and Davis came to call on Agnes and meet Russel, who is due this evening with Edwards. The boat was up to time so I went to the burra bungalow. Russel was very affable and Edwards appeared glad to see us.

Sun. Dec. 15th

Had an early *chota* and then drove with Davis to the polo ground for a parade. There was a big turnout, including the adjutant, Lawes. We did the usual work and also a cavalry charge down the ground, etc., after which an excellent tiffin was given by Lawes, the arrangements being made by Crutwell. Mears gave us a lecture and at 3.15 p.m. I made my departure. Dunlop came up for a peg in the evening and after he had been there a short time Russel turned up and was introduced. He is as sweet as honey this visit and agrees to all our proposals.

Wed. Dec. 18th

Russel left and Dunlop drove Agnes to polo.

Wed. Dec. 25th

Left for Borpukhuri at 11 a.m. via the government road, as Agnes had not been that way. Edwards and Dunlop were also at breakfast and in the afternoon Duguid and Crichton, Crutwell, Burton and Haines arrived. Tennis was quite good and afterwards we had bridge till dinner, at which Banyard, the Pabhoi manager, made a twelfth. He is a fairly tall and dark man with a big moustache. Macrae took in Agnes and Edwards sat on her right. There was a piece of genuine holly for the pudding and we wore paper caps from the crackers. After dinner were games, such as the thimble race and potato race, then a lot of singing, including 'Auld Lang Syne'. The buggies were ordered for 12.30 a.m. It was about the nicest Christmas I have had.

Thurs. Dec. 26th

Had *chota* about 9.30 a.m. and at 11.30 we left for breakfast at Crutwell's and called on the Smiths. After breakfast we played bridge and returned home about 5.30 p.m.

Sat. Dec. 28th

Went round everywhere and during the evening Johnny rode over, but did not stay.

Wed. Jan. 1st, 1908

Drove to polo. Twelve men turned up but it was rather a dull meet on the whole and we weren't late in leaving.

Sun. Jan. 5th

The doctor came over to see a case of supposed smallpox and vaccinated Agnes and myself. Dunlop returned at night from Dibrugarh, where he had been to see the dentist.

Mon./Tues. Jan. 6th/7th

Dunlop went round Monabari and Kationabari with me and appeared satisfied.

Sun. Jan. 12th

Dr Smith came over and looked at our vaccinations. At 3 p.m. we left for Butcher's. He has hurt his knee and was only just able to limp about. We got back at 6.30 p.m.

Wed. Jan 15th

Agnes and I breakfasted at Gingia, where Major Halliday was staying; then to polo where the usual crowd turned up, barring the Macraes and Butcher. Cooper is dead and Mrs Mac. is looking after Mrs Cooper.

Sun. Jan. 19th

A cold morning and the fog lasted till 11 a.m. Davidson came to call but returned to Bargang for breakfast, taking Dunlop with him. Drove Agnes to see the haunted *bheel*.

Tues. Jan. 21st

Dey and Davis came over and played croquet. I started surveying both boundary grants at Kationabari.

Fri. Jan. 24th

The doctor came over to see Agnes.

Tues. Jan. 28th

Agnes and I went to breakfast at Chota Pukhuri and left at 3.30 p.m., as a storm appeared to be coming up. It caught us just after passing Crutwell's bungalow and proved to be the worst hailstorm I have ever experienced. In addition to the fall of stones about as big as pullet eggs, great lumps of half-melted ice came down, mixed with torrents of rain, while the thunder and lightning were incessant. The Bishop behaved splendidly, though his nerves were sorely tried when we passed Kolapani, where the hail on the tin roofs made a deafening roar and the screens from the rubber trees were blowing about the road. We turned into the polo ground and sheltered under the pavilion till the worst was over. Dunlop had sent Lettice to meet us and we picked her up near the dismantled kyah's shop, where the wind had lifted up and bent the tin. Dunlop came to dinner and was very affable.

Wed. Jan. 29th

Dunlop drove his four-wheeler to Gingia for tiffin and on to the polo ground. We didn't go. Loafed during the morning and at 4 p.m. drove Agnes to Bargang. We found D. at home, recovering from a dose of fever. We had tea and returned about 6 p.m. He presented us with a very pretty mantle border, which was decidedly generous of him.

Wed. Feb. 5th

Drove Agnes to polo, where there was quite a big meet. Swinley proposed the Robinsons' health, as it was their last day. Then D. proposed Swinley's. He made a good bull by saying that he hoped Mr and Mrs Swinley would continue to run the ground from time 'immemorial'. D. left early

184

as he had got fever. He drove Agnes back in the four-wheeler.

Fri. Feb. 7th

Dunlop inspected Kationabari and didn't altogether like the pruning.

Mon. Feb. 10th

Dunlop and I first went to Kationabari where the pruning was bad; then at 11.30 a.m., left for Bargang to try the new road, which we found quite passable for a four-wheeler. At Bargang we met Cattell, the new assistant. He is a man about a year or two older than Dunlop. He comes from Sylhet and told us about work on bheel gardens. After breakfast I biked to where our wood cutters are working.

Wed. Feb. 12th

Wet day. Started pay and in consequence couldn't go to polo. Took pathar rents.

Sat. Feb. 15th

D. wrote a decidedly impertinent letter re the Bishop. His facts, as usual, were wrong. When starting for Kationabari in the afternoon, the Kacharis at the office shifting bamboos, frightened the Bishop and he took the buggy into the tea and broke the axle, so I walked to Kationabari and back. Braddan called and the Whytes' PWD came to stay with Dunlop. We dined, including Braddan.

Sun. Feb. 16th

D. and Whyte surveyed the river on the Behali *hathi*, but the PWD won't have anything to do with our drain. Agnes played croquet with Mrs Whyte, who is not much good. Mr Beatson Bell, his sister and Mr Sweeney, the land settlement people, rode over and had tiffin with the whole party, leaving shortly after.

Tues. Feb. 18th

Dey, Davis and Dunlop dined with us. The table looked very pretty and everything went off first class.

Thurs. Feb. 20th

Johnny left to take charge of Kootrie.

Thurs. Feb. 27th

D. left for a managers' meeting at Majulighur, while I took Antram out shooting behind the bungalow. We put up a lot of deer and he got a couple, both shots being good. He also bagged a partridge. He breakfasted with us and at 5 p.m. left for Bargang. The Bishop played the fool coming home and went off the road, but J.J. and D. were driving past and lent a hand.

Sat. Feb. 29th

Went to visit the drain where the Kacharies are making a bridge over the Bargang for the women to cross while making the bund. Agnes and I went to Bargang for the night and at the bungalow we found Lieutenant Lyall of the military police at Dibrugarh, who had come to inspect the guard. I recognised him as the man who came to meet his fiancée, Miss Iles, on the *Syria*. We had a talk and he said his wife had been keeping well. He invited us to go and see them.

Sun. Mar. 1st

Lyall went to inspect the guard while D. and I set out for the range. I made 174. Lyall and Agnes, followed by Butcher, arrived when we had finished at 700 yards. Lyall had some shots and then took Agnes back, while Butcher drove D. and myself in his four-wheeler. D. has brought a fine bath from home so I got the best wash since the Grand Hotel in Calcutta. After breakfast I went to the forest and on my return found Lyall had gone. We left about 5 p.m., and reaching home I went to the burra bungalow as Mitchell, who had been surveying all day, had arrived.

Mon. Mar. 2nd

Went round with Mitchell, surveying the N.L. grant. It is much simpler work than I thought. We had tiffin on the work and finished at 5 p.m.

Wed. Mar. 4th

Started women making a bund opposite our drain. Drove Agnes to polo. She was the only lady there.

Sat. Mar. 7th

Dunlop hears a new assistant, Cameron Douglas, is to arrive about April 1st.

Tues. Mar. 10th

First of wearing white clothes.

Wed. Mar. 11th

Mounted parade at Kolapani, the sergeant-major and Captain Lawes being in attendance. I drove the Bishop and also drilled him. There were twelve on parade. Dunlop drove Agnes for tiffin, which was supplied by him. Afterwards the usual tedious period happily enlivened by the gramophone. Then, at 4 p.m. Crutwell drove Mrs Smith up while the doctor brought Ferguson, a Johat man. The arrival of Butcher completed the company. There were three chukkers and shortly after they were over Agnes and I left.

Thurs. Mar. 12th

Today we decided to completely bund the Bargang and send all the water down the drain. As Agnes wanted to see it I drove her to the seed garden and we walked from there. Dunlop arrived at the same time via the brick kiln and we commenced by putting down bearers and thatch. In less than an hour the job was completed with nothing to do but strengthen with earth. Godwin arrived about 6.30 on a short visit.

Fri. Mar. 13th

Dunlop took Godwin to see the drain and at noon they came up and called before going to Gingia. Godwin invited us to Addabari and we are to return with him on Tuesday.

Sat. Mar. 14th

Dunlop and Godwin went to Borbheel for the night. I had to pay all the coolies as the money had only just arrived. It took me from 2.15p.m. till 7.15 p.m. and I knocked off about half an hour for tea. Altogether I paid nearly 1700 souls.

Mon. Mar. 16th

Dunlop took Godwin round Kationabari and they dined with us. We all turned in early as we are to return with Godwin to Addabari tomorrow.

Tues. Mar. 17th

Left at 7.45 a.m. Godwin drove Agnes in the four-wheeler and I went in the big buggy. We had to wait an hour for the steamer. The boat was very leisurely in going down and we reached Tezpur just as the train was due to start. However, we caught it and found Mrs Penman and Cragoe were travelling with us. Mrs Penman got out at Bindukri and Cragoe at Thakurbari, as he is going to Bamjarm. We three went in one buggy (I doing syce) to Kacharijan. I had not been there for five years and the bungalow has been greatly enlarged. A great number of men were present: Hannen, Percy, Will Briscoe, Allen, Causton, Moran, Playfair, etc. We only stayed a short time and reached Addabari around 7 p.m. Godwin's new assistant, Mungo or 'Mango' Smith, is living with him. He is a small, nippy little chap, just like an assistant in a draper's shop. We played bridge after dinner.

Wed. Mar. 18th

Godwin's tennis day. Mortimer, Chaston, Drake, Dr and Mrs Patterson, Arthur Moore, Curtis, etc. turned up. Two courts were kept going.

Thurs. Mar. 19th

Went to the new garden here and admired the terracing. About twenty men were present at polo. After dinner we dined at the Pattersons. Causton, Archie Hannay and Reid were also there.

Fri. Mar. 20th

Went to breakfast with Chaston, who is living in the burra bungalow. It is a fine building with stained walls in the drawing room and a wainscot. We amused ourselves with target shooting and then I went to see my car at Baxter's bungalow. The car looks A1, but wasn't running, owing to a small gear wheel not having arrived. Then we went to the tea house, which is beautifully arranged, with any number of labour-saving devices. Afterwards breakfast; then Chaston took me on a drive round the garden, the whole being 1500 acres, one clearance – seed at stake – two years' old, being magnificent. Their *puja* started today. A few people came to tennis, including Godwin, and we three drove back together.

Sun. Mar. 22nd

Loafed in the morning and went with Godwin to Phoolbari tennis in the afternoon. Agnes was seedy. We were not late returning.

Mon. Mar. 23rd

Walked to the new garden in the morning and after breakfast Godwin drove Agnes to polo, as before, while Walter took me. There was no news, except that Hutchi of Gillahating has gone to Kassuli and on his way to Tezpur his syce let his pony and buggy fall in the Borelli. The pony was drowned. We went by evening train to Tezpur and it was after dark when we reached there. As the boat wasn't in we went to the chummery. Cragoe gave us dinner and we lay down for a spell.

Tues. Mar. 24th

Got on board at 3 a.m. Turned Baxter out to make room for Agnes. Cold windy day and we stayed in the cabin. Reached Behalimukh at 2 p.m. and found Lettice and the four-wheeler. There has been a little rain, which was the first during this month. On arrival went to see Dunlop and he told me the *puja* had been very quiet and everybody went to work on Monday, the only misfortune being that the larger half of the nurseries got burnt, including the Monabari seed.

Mon. Mar. 25th

We plucked yesterday about sixteen maunds. Went today with Dunlop to Kationabari where the unpruned is flushing. He went to polo but we did not.

Sat. Mar. 28th

Dr Smith's farewell tiffin at Partabghur. Agnes and I did not go because it is such a long drive in the heat of the day.

Sun. Mar. 29th

We loafed all day and in the evening went to the burra
bungalow to hear about the tiffin, which was rather spoilt
by Biddy having fever. Mrs Mac. first stayed with her and
after tiffin Mrs Smith went back.

Tues. Mar. 31st

A regular heat wave seems to have set in. The thermometer
rose to 97°, which is the second highest I have ever seen.
Only twice have I seen it at 98°. We started breaking land
for the clearance.

Wed. Apr. 1st

Another hot day, this time 96°. We went to polo and said
goodbye to the Smiths. A regular dust storm blew hard
and during the night about half an inch of rain fell.

Mon. Apr. 6th

Got a note from Dunlop, saying that Cameron Douglas is
due today and asking me to go and meet him. I reached
the ghat about half an hour before the steamer, which was
due at 11 a.m. Douglas rather resembles Godwin, tall, dark
and wears a pince-nez. He told me his isn't much of an
engineer, having only been in a tool shop, but also managed
a cardboard box factory. He looks about twenty eight.

Wed. Apr. 8th

D. came over for a shoot and joined me at Kationabari,
where we mounted the *hathi* and went to the new brickfield

bheel, which we crossed, and kept westward until we came to another round *bheel*, not large but full of duck and teal. Shortly after we came to another, where someone had built a hut. Then we sighted the long *bheel* (which runs to the Gingia *nadi*) but turned south west to some low grass country. We didn't enter it but kept south over some high ground, where we came on rhino tracks. These led to some very heavy jungle, which we tried, but it was too much for the *hathi*. We gave it up and returned home, reaching the bungalow at 3 p.m.

Sun. Apr. 12th

There was a service at Borpukhri at 10.30 a.m. I drove the Bishop to Kolapani and got the Macraes' pony there while Dunlop drove Agnes. Edwards, his new assistant, King, Dr McCombie – very boyish-looking – Banyard, Crutch, Davis, Burton, Duguid, Crichton and ourselves were present. The padre is a pale and rather emaciated-looking little man. During the afternoon we had tennis and got home at 8 p.m., only to find that our cookhouse has been burnt to the ground, so we had to dine with Douglas. D. had been shikarring and Butcher had ridden over to call. They helped to put the fire out.

Mon. Apr. 13th

The DSP, Plowden, arrived to inspect the pound about 7.30 a.m. He is a huge man who has been eighteen years in Burma. He spent the day with us and I took him to see the staking of the clearance started. Dunlop returned at 7 p.m. and Plowden stayed the night with him.

Tues. Apr. 14th

D. came over with his *hathi* to search for his keys, which he dropped on Sunday. I went out with him and he told me he had met Woods and together they went on foot into some rhino tunnels and that Woods had seen a rhino and shot at it. We went to the duck *bheel* and from there to the grass country, where in one place we found a tunnel, and dismounting, searched for the keys, together with some of his coolies. I had never seen a tunnel before and it was very interesting. There was scarcely room to stand up and it would have been almost impossible to force a way through the sides. From here we went to a small open space where the torn-up grass showed that rhino had been feeding there a few hours previously. A long tunnel led out of this and we sent the coolies into it and took the *hathi* the most convenient way. The tunnel was about half a mile long and emerged on to high land. We went across this and finished in some heavy stuff, pierced with tunnels, where D. showed me the marks of Woods' shots. He must have marvellous nerve, as where he fired he was only a few yards away. We couldn't find the keys anythere, so returned to the bungalow at 5 p.m., very thirsty and dirty. Plowden came up and called about 7 p.m. Then he and D. went to the burra bungalow, as they were staying the night.

Wed. Apr. 15th

Had a row with Dunlop. He is getting unbearable.

Sat. Apr. 18th

Mrs Macrae spent the morning with Agnes and they both had breakfast with us, while Edwards, Davis and Butcher called and then had breakfast at the burra bungalow. I drove to Kationabari to try for a buffalo which Woods had wounded in the leg but couldn't find it.

Mon. Apr. 20th

Got word that the jungli buffalo had been found so I went after it. We crossed the *chur* next to the old brickfield, going south east, till we reached the Brahmaputra. The buffalo was in the water some distance away so I waded in for a bit and fired about four shots. The result was two hits and two misses. I knocked it down but it didn't seem to have died. However, it was now nearly dark so I returned.

Wed. Apr. 22nd

Drove to Gingia for breakfast. The Macraes and Dunlop arrived at 4 p.m. I drove down to Kationabari and crossed the *chur* to where the boat was to be. It wasn't there and as D. just then turned up we decided to try the jungle. We came on a buffalo in a little *pukhuri* due south of the jute but D. missed him and as it was getting dark we gave it up. D. stayed the night and Douglas dined with us.

Fri. Apr. 24th

A slight rainfall but we started at 6 a.m. We tried round the *pukhuri* but could find nothing so at 8.30 a.m. I gave it up and returned, having work to do. D. went on after rhino and returned about 2 p.m., having seen one, but as the *hathi* bolted he didn't get a shot. He left for Bargang at 3.30 p.m.

Sat. Apr. 25th

Dunlop returned from Mijica, bringing Dr McCombie, who went to see Agnes.

Mon. Apr. 27th

Dunlop started transplanting. Douglas has got a bad foot and is confined to the bungalow.

Thurs. Apr. 30th

Started transplanting in a big way with ten carts and all women from the lines.

Tues. May 5th

Three point seventy five inches of rain, so transplanting was impossible.

Wed. May 6th

All hands on the clearance. Agnes and I went to polo. Miss Robinson and Mrs Simmonds were there. Butcher invited himself for breakfast on Friday.

Fri. May 8th

Douglas seedy. First packing of the year. Butcher arrived at 12.30 p.m. and after breakfast I took him round the clearance. He went at 5.30 p.m. and I drove Agnes to Kationabari.

Thurs. May 14th

Inspected Kationabari before pay and drove out to the government road. The mohurir and I got out of the buggy to look at the bamboo bari and while Hirasai was collecting the reins the Bishop bolted and galloped straight to the river but no one was injured. Hirasai took him back to the stable and, changing the buggy for the saddle, brought him down for me. Dunlop had tea with Agnes and stayed till 8.15 p.m. I thought we should have to change dinner into supper.

Fri. May 15th

At 3.30 p.m. we started for Mijica. At the hat we got Kathleen and reached Mijica just before 5 p.m. Only Davis was there so we had a nice four. We left about 7 p.m., and had a glorious moonlit drive back.

Fri. May 22nd

Dunlop went to Mijica for the night. The sub deputy controller surveyed the new grants and passed everything.

Sat. May 23rd

Dunlop returned with Dr McCombie. Douglas's foot is bad again. The DB at Kationabari had another row with the coolies.

Sun. May 24th

Dunlop interviewed the DB who lost his head and was insolent, getting the sack in consequence, but after a written apology Dunlop decided to retain him. But in future I am to give more attention to Kationabari, having my chota at 6 a.m., and working at Kationabari till about 10 a.m., and also trying to be in attendance at noon and 6 p.m.

Fri. May 29th

Major Woods and two *hathis* arrived for a shoot. Also D. arrived.

Sat. May 30th

D. and Woods made an early start but saw nothing though they didn't return till nearly 7.30 p.m. All women on the clearance and we now have over fifty acres planted. The Bishop bolted from his two syces near the stable and smashed the big buggy to pieces. He is now too dangerous to drive.

Sun. May 31st

At 4.15 p.m. Agnes and I went to the burra bungalow to say goodbye to Dunlop and get the four-wheeler and Lettice. We found Woods and D. there. They had been absolutely chivvied out of the *bheels* by at least four rhino. The *hathis* kept bolting and they couldn't get a shot. We left at 4.30 p.m. with the hood up as it was very hot, and at Kolapani found Flashlight with the Macraes' buggy. At Mijica we met Burton, who said Crutwell was at the station. We called at Chota Pukhri, finding both the doctor and Mrs Simmonds and reached Borpukhri about 7.15 p.m. Only the Macraes were there and we had dinner, played one rubber of bridge and then went to bed.

Mon. June 1st

Agnes, Mac. and I had chota at 8 a.m. and at 9 a.m. I left, driving Flashlight to Kolapani, where I found the Bishop in the saddle. I looked in at the burra bungalow, where Woods and D. were having a drink, preparatory to the former's departure; then on to the tea house to work at the engine, I spent the afternoon there.

Thurs. June 4th

The Imperial Tea Company report arrived, showing our profit of £5000, which is nearly £1000 too little when the sales are checked against the gross expenses.

Fri. June 5th

Dunlop drove me to Mijica, where we only found Davis. But after a little while Mrs Macrae drove in with Agnes, followed by Macrae and Mrs Simmonds. Tennis was quite good and about 6.30 p.m. Agnes and I left, using Flashlight to the hat and Kathleen from there.

Tues. June 9th

Dunlop got a wire, saying, 'Allanson dead. Transfer H. to take management.' I went up at 3 p.m. to discuss the matter, which is not very pleasing to either, as half of the commission here next year would equal three years of Ruthna, which is a rotten show. On his part Dunlop doesn't know where to look for an assistant. So he wired Calcutta if he could keep me till November.

Wed. June 10th

Heavy showers so no one went to polo. Dunlop got answer that I must be transferred at once.

Thurs. June 11th

Agnes' birthday and my present to her of some new tea things arrived safely. Wired McLeod, 'Is billet permanent?' and got answer, 'Do not remove effects.'

Fri. June 12th

Douglas developed fever and his temperature reached 106° at 3 p.m. I sent for the doctor who arrived at 8.30 p.m. His temperature was then 105° but when I went home it had fallen to 103°.

Sat. June 13th

McCombie turned up during the night, having ridden the Bishop, who bolted with him from Kolapani. He didn't seem to think Douglas was bad and left at 11 a.m. We spent the day packing.

Sun. June 14th

We are taking twenty two packages, all told, but that doesn't include any furniture. We breakfasted at the burra bungalow, where D., Davis and Butcher arrived to give us a farewell. Dunlop was rather silent and never proposed our health or said anything with reference to the period of my assistanceship. He merely seemed to resent my transfer. We left in the little buggy with Lettice, who took us to

Kolapani, where the Macraes had sent a dak. We found no one at Mijica and didn't call at Chota Pukhuri. At Borpukhri Crutwell was waiting to say goodbye and soon cleared off. We had dinner at 8 p.m. and went to bed immediately after.

Mon. June 15th

Walked round with Mac. Edwards came to breakfast in his silkiest mood and stayed till 5 p.m. When we left the Macraes gave us a much more hearty send off than Dunlop. We had a lovely drive to Bishnauth and arrived at the same time as the steamer. We had to boat to the flat, which was lying downsteam on a *chur*. The boat was a new one, the *Mirani*, with electric fans on board both in the saloon and cabins. We spent the night at the ghat.

Tues. June 16th

A cool drizzly morning. I turned out as we came to Silghat; it was like my first trip down in 1901. Had chota at Tezpur, where we stayed till 10.30 p.m. The rest of the journey was uneventful, save that the evening was lovely and Gauhati looking almost at its best. I went straight to the station and found the train was just going so returned to the boat for dinner. Left our things in charge of the *khansama* at the flat, and went to the dak bungalow.

Wed. June 17th

Went to the ghat and brought the luggage by the branch line to the station. Sent a note to Major Herbert, who asked us to tea, as it was too late for breakfast. His bungalow is high up over the river with a lovely views. The Herberts are as nice as ever and were very pleased to see us. We got the train at 6.20 p.m. and dozed till we reached Lumding at 1.30 a.m. There was no trouble changing and I was also able to get some tea.

9

RUTHNA TEA ESTATE

Thurs. June 18th

Stopped at Haflong for chota at 9 a.m. We are in the hill
section, which is very like the route to Kandy, only there
are bamboo forests everywhere. The rail runs through the
Jitinga valley and at noon we came out on to the plain and
crossed the Surma by a fine bridge and pulled up at
Bardaphur junction for Silchar. We breakfasted there and
restarted about 2.30 p.m.

At 5.40 p.m. we stopped at Juri siding and found a
sirdar, ten coolies, a four wheel buggy and pony waiting.
The road was simply awful in some places and absolutely
dangerous, being narrow and slippery. The bridges were
bamboo mostly and one corrugated iron with no *matti* on
it. We passed through one garden, Kapnapahar, then
through a cutting in a *tillah* and came onto more *bheel*
country with Ruthna tea house and bungalow in full view.
The latter was up on a *tillah*. We had to get out a few
hundred yards from the bungalow and walk up the steep
hill, on the top of which it was perched. Mr Lees was
waiting for us. He is probably fifty years old and seems a
decent sort. The bungalow has a large verandah but the
inside is more like a tumbledown barn than anything else.
The posts are wood and where the original ones have
rotted extra ones have been put in with no regard to the
appearance of the room. There are three rooms altogether,
but absolutely no furniture, and everything is covered with

dirt, the pucca floors broken and with no carpets. Agnes was in despair and said she would not stay, but the next day she thought it was not so bad.

Fri. June 19th

It drizzled in the morning and the bell was not rung, as the coolies do not turn out when it rains at all heavily. Lees had slight fever, so we took it easy, but late in the afternoon we went into the garden. It is a very straggly affair, about three miles long, and intersected by *tillah* and *nullah*. The southern half of the garden is known as Ellapore.

Have now passed one full day, so the good and bad aspects of the place can be summed up. The bungalow is bad but the view from it is beautiful. The machinery is short, being only one 'rapid', one 'cross-action', one 'down-draught', one 'sixteen side tray' and another wee drier with about four trays. The tea house is a fine building, however, but with only one loft. A lot of the tea is good but there are fifteen per cent vacancies. The labour force is nearly two coolies per acre, but fully a third have no agreements. Allanson's pay was only Rs.400 per month but three years ago the garden made a loss. This year the profit has risen to over £700.

Sat. June 20th

Another drizzly morning, so the bell did not go until nearly 8 a.m. I paid tickets for the first time. They are paid here every two days and the system is as follows: each large ticket represents a full task for which three annas is paid. The small tickets are quarters of the large; by this a child who does a full task gets paid as much as a man, but their tasks are arranged so that they seldom get full pay. This system has the advantage that a coolie gets paid immediately after doing his work. The coolies are very orderly at

204

the pay table and make far less noise than they do at Monabari. Lees and I worked in the office till 12 o'clock and he went away in the evening.

Sun. June 21st

Today is not a leave day here as the *hat* is held on Tuesday. During the afternoon four men came to call: Harrison, manager of Clevedon, and Sandeman, his assistant, Ekins, the manager of Silloah and Kennedy, the district engineer. Mrs Heathcote of Kapnapahar sent over a chair slung on bamboos for Agnes, who went across in it. It is only two miles. They seem nice, friendly people.

Mon. June 22nd

Went round part of the garden. It is very hard to get about, owing to lack of roads and a great number of drains. The acreage is 365 plus fifty clearance. A lot of the tea is very good, but some plots simply awful, especially the far end of Ellapore, which got burnt last cold weather. The establishment consists of four babus, a DB, rather second rate, a KB, who is a cheeky young blood but smart, a Mussulman who looks after the women and another Bengali babu, who runs the men. I had the last three up and warned them that if any trouble arose they would be the first to go.

Tues. June 23rd

Our leave day but the tea house was working, as they have been round in eight days. The Heathcotes came over in the afternoon, both riding. She is a very diminutive person, rather elfish, while he is about my height. We talked a lot of shop and he told me he made about £3000 profit off just over 300 acres. He plucks coarse and gets an outturn of ten maunds per acre.

Thurs. June 25th

My thirtieth birthday and I don't feel more than twenty five. I went round the garden in the morning but stayed at home during the afternoon.

Fri. June 26th

Our medico, Dr West, came to call. He is about thirty five but only came to India last February and so finds things rather strange. Lees called. He has arranged for a pony rickshaw for me to buy if I feel so disposed. Price Rs. 120 for pony, harness and rickshaw, which is very cheap.

Sat. June 27th

Lees and I went into some office matters and at noon he left again.

Sun. June 28th

Etkins having lent us his rickshaw, I ordered four coolies to pull it and took Agnes over to Kapnapahar, myself riding. We found the Heathcotes at home. Their bungalow is very small and the eaves terribly low, so that one has to stoop to enter. No one seems to have fireplaces in this district and everything is most kutcha.

Tues. June 30th

Have started jungle cutting on the clearances. The bushes are looking splendid. Our leave day but my time in the morning was chiefly taken up in paying people. In future I shall do this on Monday evening.

Thurs. July 2nd

The doctor breakfasted with us. I have got a bad go of rheumatism in my neck, feet and right hip bone, which hinders my work.

Mon. July 6th

Steady rainfall day. Work poor and short in quantity. We have hardly seen the sun for a week and the temperature has seldom been over 85°. Leaf growing very slowly.

Wed. July 8th

Felt seedy but rode to Montijum to see plucking and hoeing. Fever came on so I returned quickly. Temperature 103°. However, after fifteen grains of quinine it fell to 100°. Kennedy called.

Thurs. July 9th

Dr West came to see me during the morning and went on to Clevedon for the night.

Sat. July 11th

Rheumatism bad all over; am scarcely able to walk.

Sun. July 12th

Perpetual rain; everything damp and cold.

Mon. June 13th

Agnes got fever for the first time.

Tues. July 14th

Agnes' temperature normal at daybreak but over 100° at midday, so I sent for the doctor. He arrived at 8 p.m. in a heavy storm and stayed the night.

Wed. July 15th

During the afternoon I went to Silloah with the doctor as far as Kapnapahar. I rode and he walked but at the tea house we met Heathcote and West borrowed a pony from him as far as the ghat. The road through the lines was disgraceful but the ghat is only about half a mile away. The Juri is a small stream with steep permanent banks. We crossed in a mah. The road on the other side is rather pretty and Etkin's four-wheel buggy met us after a short distance.

Silloah garden seems to be much higher than Kapnapahar or Ruthna. The bungalow is small, three rooms, but very trim. The Etkins were in the verandah. She is a small person and pretty in a way. Her eyes are blue grey and at times look very hard. I expect she rules him. I stayed till about 6 p.m., then drove his buggy to the ferry and rode home.

Thurs. July 16th

Agnes feeling better, though rather weak.

Fri. July 17th

Rode to Clevedon in the afternoon. The road through the range is rather pretty but the view is confined, with dense bamboo jungle everywhere. The defile suddenly emerges without warning onto the Clevedon garden, which is mostly *bheel*. The bungalow is high up on the opposite side, the tea house being at the foot of the *tillah*. I was able to ride the whole way up the path leading to the tennis lawn, where I found the Heathcotes, Etkins, the doctor and Jack Barry. The tennis was rather poor.

Sun. July 19th

The piano arrived. The doctor came to breakfast and was very pleased to get some music. He intended to leave us about 5 p.m. but as a storm was brewing he decided to stay and was walking with me to the leaf weighing when we met Mrs Etkin in a chair with Etkin and a tall, dark young man called Montagon walking behind. We all got to the bungalow just as the storm broke. Half an hour later Mrs Heathcote walked up, so we had a big party. Montagon has a good voice and sang 'The Devout Lover' and 'Songs of Araby', while I obliged with 'Take a Pair of Sparkling Eyes', etc. They left about 7.30 p.m., when it cleared a little.

209

Mon. July 20th

My rheumatism seems worse than ever. The doctor has given me some strong medicine.

Fri. July 24th

Getting good leaf now. Today fifty one maunds.

Mon. July 27th

Agnes got a dose of fever.

Thurs. July 30th

The doctor came to breakfast on his way to Rajki, where they have cholera.

Mon. Aug. 3rd

Lees coming on Tuesday. All women on the plucking. Only twelve maunds.

Tues. Aug. 4th

Lees arrived midday but before we expected him, as our clocks were slow. He has got boils and is quite seedy. We took a little walk but I got low fever so we were not energetic.

Wed. Aug. 5th

Went round with Lees. He appeared satisfied and said the

weather must be in fault and not hot enough. We returned at noon, when my temperature was 100° and didn't go out again.

Fri. Aug. 7th

Lees announced his intention of leaving directly after *chota*, which he did, riding Rodney to Juri.

Sun. Aug. 9th

Mrs Heathcote came over in the evening. My fever has been continuous.

Tues. Aug. 11th

Fever normal during the early mornings but always 99° or over at midday. Quinine has no effect.

Wed. Aug. 12th

The DB has put me on a diet and has stopped me from going out to Kamjari. Sandeman came over and stayed an endless time.

Fri. Aug. 14th

The doctor came and, finding that my ankle was swollen, ordered me to bed, to drink nothing but milk and take very strong medicine.

Wed. Aug. 19th

The doctor gave me permission to go out. Went round the garden in the rickshaw. It looks fairly healthy and more leaf than when I was last out.

Thurs. Aug. 20th

Went for a ride but only putting one foot in the stirrup.

Fri. Aug. 21st

Ankle has swollen up again, so am once more confined to the bungalow. Have picked up a good deal in crop this week and am now only 200 maunds behind.

Tues. Aug. 25th

The Heathcotes came to lunch; also the doctor, who ordered me a daily hot water soak for two hours in the morning and two at night. They all went away at 5 p.m. and it was a relief, as somehow they were tiring.

Thurs. Aug. 27th

Sent the rickshaw for Nurse Vanguland. Sandeman had the cheek to send his things over so I wrote a chit, returning them as he emerged from the *phari* he rode on to Kapnapahar. Nurse arrived about 6 p.m. She seems quiet and capable.

Fri. Aug. 28th

The doctor rode over from Kapnapahar but returned there for breakfast. He is arranging to live here from Monday.

Sat. Aug. 29th

Had a chair rigged up and was carried to the tea house. Returned at 11 a.m. when nurse said I must send for the doctor. Agnes was feeling well and had breakfast with us and also afternoon tea, at which time the doctor turned up and thought he had been sent for unnecessarily, as she was feeling well at dinner time. Nurse slept with her, the doctor and I in the spare room. About 11 p.m. I heard nurse calling the cook for hot water and shortly after she came and roused the doctor. He returned with her and on coming back said all was well and he expected the birth about 6 a.m. The servants were present, so we had some tea and I followed with a strong peg.

Sun. Aug. 30th

About 2.30 a.m. nurse surprised me by bringing news that it would be over in twenty minutes, so I waited anxiously and at length heard the first cry, which seemed very powerful; then the doctor's voice saying, 'Well, if you want to know, it is a little girl.' A few minutes after the nurse came in to congratulate me and finally the doctor came and gave some particulars. He said she weighed eight pounds, which was good, was a 'ripping little child and had a lot of dark hair.' He had a peg and shortly after turned in. Baby was born at 2.55 a.m. We were all late in rising and then I was allowed to see Agnes and the baby.

Mon. Aug. 31st

Agnes and baby both well. I was allowed to sit a long while with Agnes and also see the baby bathed.

Wed. Sept. 2nd

We are now only 170 maunds behind and prospects are good.

Thurs. Sept. 3rd

The doctor nowadays makes short calls during the day to Clevedon, Kapnapahar, etc., but always returns at night.

Fri. Sept. 4th

Agnes got a dose of fever up to 103°, but it broke during the evening and dropped to 100°.

Sun. Sept. 6th

Have started to walk a little, but my foot is very weak.

Thurs. Sept. 10th

Am now plucking finer and leaf has fallen off in quantity.

Fri. Sept. 11th

The doctor left for Silloah and has decided that he need not return any more, except for casual visits. Agnes' bed was wheeled into the verandah.

Thurs. Sept. 15th

Have decided to go to Surma on Thursday for a couple of nights.

Thurs. Sept. 17th

Left about 10 a.m and at Juri had forty minutes to wait. Got breakfast in the dining car. It was hot at first and not a very comfortable journey. Reached Surma siding about 5.30 p.m. A four-wheeler was waiting and after half a mile or so got into tea and another mile pulled up at the bungalow, where Lees was waiting. The bungalow is two storey with six rooms and brick built walls to the roof and a pukka fireplace and chimney like a home house. The roof is corrugated iron. Lees and I had a good talk, then dinner and an early bed. I feel better already for coming.

Fri. Sept. 18th

A wet morning so we loafed in the verandah till it cleared about 10.30 a.m., then took a little drive round, but the roads don't extend far. The garden has a sandy soil on which jungle doesn't grow at all. At 6 p.m. we drove to Teliapara garden about one and a half miles away, for dinner with Jones, who is in temporary charge with a young assistant, Dennie.

Sat. Sept. 19th

Rose before daylight and got the train about 6 a.m. Soon after had chota and then very nearly got a fit of ague and was glad to get in the hot sun at Juri. A storm caught me just before the tea house and when I reached the bungalow

215

I was dead beat. Agnes and baby were not very fit so things were not cheerful. Went to bed with a temperature of 104°.

Sun. Sept. 20th

The doctor called and went on to Clevedon as Sandeman had got a bad ear. Nurse wants Agnes to go to Shillong with her and as she needs a holiday I may send her.

Mon. Sept. 21st

Worked in the office and then rode round the garden. There is absolutely no leaf or bud either, but today is sunny and ought to help matters.

Sat. Sept. 26th

Decided not to send Agnes to Shillong. Nurse left in the afternoon, taking the train to Kalaura and waiting there for the up train at midnight. Agnes has asked Mrs Heathcote to come and stay for a few days.

Sun. Sept. 27th

Mrs Heathcote was to have come early but we got a chit saying she had an accident and would not be over till the evening, so Agnes had to bath the baby herself and was very nervous. Mrs Heathcote arrived at 5 p.m. and said she had been thrown from the trolley, which had jumped the rails.

Wed. Sept. 30th

Heavy rain at last, a total of two and three quarter inches. Agnes got a dose of high fever, temperature 104° and palpitations during the evening. Kennedy and the doctor were here.

Thurs. Oct. 1st

Leaf is breaking away and we got over forty maunds. Got two wires, one saying a Canning House nurse arrives Sunday morning and the other from Edwards, saying a nurse for baby arrives, also on Sunday.

Sat. Oct. 3rd

Sent a rickshaw and chair overnight to meet the nurses. The *puja* is opening very quietly. Sandeman called but didn't stay long.

Sun. Oct. 4th

The nurses arrived about 6.30 a.m. Nurse Burnell is stout and good-humoured while baby's nurse is a slim Eurasion girl of twenty. I have given her the bungalow office as her room for meals and sleeping. Mrs Heathcote left about midday, after having earned our undying gratitude. Agnes did not get up and it is a relief to have a nurse for her.

Tues. Oct. 6th

Everyone turned out to pluck and we got fifty maunds. The *puja* has gone off very quietly.

Sun. Oct. 11th

Lees asked if he could come over and arrived at breakfast time. We took a walk round in the evening, visiting the clearance and returning by Ellapore lines. This is the most walking I have done. Things are going more smoothly with the two nurses here.

Mon. Oct. 12th

Lees and I went as far as Montri Joom and he left by the Assam mail. He is very keen on my taking a holiday.

Tues. Oct. 13th

The doctor arrived, very apologetic after his visit to Lungia. He stayed to breakfast and I rode with him to Clevedon. Just as I was leaving an urgent message came, saying baby was unwell. So the doctor followed me. It seems she had a sort of convulsion but was better when we arrived.

Thurs. Oct. 15th

Agnes, nurse and baby have all got fever.

Fri. Oct. 16th

Sent to Sagurnal for the doctor to come early as he promised to do, but according to his fashion he arrived in the afternoon.

Sun. Oct. 18th

Baby's fever doesn't leave her and we are relying more on the DB than West. Nurse thinks him useless. Hear that the Heathcotes are back and write asking if Agnes can go there and get a very cordial affirmative.

Mon. Oct. 19th

About 4.30 p.m. Agnes, Nurse Burnell, baby and I left for Kapnapahar. Mrs Heathcote was delighted to see baby again. They told us they had a splendid time on the river.

Mon. Oct. 19th

Ticca plucking and got sixty eight maunds. I didn't wait to pay everyone as I was due at Kapnapahar for breakfast. Nurse Burnell left during the afternoon. Her little visit has cost Rs. 182. Nurse Boswell is now at Kapnapahar. Agnes and baby are better for the change. We have decided to go on the river, starting Thursday, and Mrs Heathcote is to look after the baby.

Wed. Oct. 21st

I have now got about forty Jali people working on the clearance for me, so it is getting on. A man was found dead in front of the DB house this morning, so I sent for the police and the body was taken to Moulvi. I gave out tickets, etc., and at 5 p.m. rode over to Kapnapahar with Sandeman, who had called.

Thurs. Oct. 22nd

Cleared up everything and at 9 a.m. rode to join Agnes at
Kapnapahar. She was carried in the chair and arrived at
Juri more than half an hour too soon. The train journey
reminded me of our trip to Darjeeling and I felt more as
if we were on our honeymoon than leaving baby behind
us. At Karimjung the rickshaw was got out and the mile
to the ghat traversed. We had to cross by ferry to the dak
bungalow, which is practically on the river bank, so I sent
the rickshaw back and ordered the men to bring up the
luggage. Agnes developed high fever and had to go to
bed, so we decided not to go on till the following evening.
This bungalow is very nice and clean, quite the best I have
seen.

Fri. Oct. 23rd

Agnes better and able to get up. The breeze off the river
was lovely and fresh and the view pretty. Just before dark
our boat, the *Ghazi*, came in so we walked down comfortably. We left at 2 a.m.

Sat. Oct. 24th

At Fenchungunge I found my tramway material and here
I sent a wire to Heathcote. A little further on, at Monoo-
mukh, we got a wire saying baby was well, which is a relief.
The country is all flooded during the rains, but now the
river has sunk and is about as broad as the one flowing
past Barisal. At night we reached Markhali where the
Surma joins the Barak and the feeder steamer was waiting
to transfer bags of potatoes.

Sun. Oct. 25th

We soon got into very broad expanses of water and at night reached Naraingunge, which is a big place and has many jute godowns. We got another wire telling of baby's health.

Mon. Oct. 26th

Reached Chandpur just after chota. The ghat lies up a sort of river and there were several sailing brigs about. It seemed strange to find sea-going boats so far from salt water. Our boat had not arrived, so we went on to meet it and finally picked it up at Barisal about 6 p.m. The journey was pretty and we were glad of the excuse for an extra day. Our new boat, the *Afridi,* left about 8 p.m. and morning found us at Chandpur.

Tues. Oct. 27th

Left Chandpur about 11 a.m. and reached Naraingunge at 4 p.m. Now a repetition of the down journey.

Thurs. Oct. 29th

At Mudna I wired to Kapnapahar and here we saw a most extraordinary man. He wore native dress but his hands and feet were whiter than mine. I was told he edits a Dacca paper. He had no look of an Eurasian about him. We reached Fenchungunge about 9 p.m. and took our things to the dak bungalow for the night.

Fri. Oct. 30th

My coolies were in attendance and I dispatched the luggage

after hammering Robbia for breaking a key and not confessing it. We reached the station about ten minutes before the train and the journey was uneventful. At Juri I got my mail and on the way to the garden Somaroo syce told me that the KB had a row this morning with Gurudoyal sirdar and his wife.

I left Agnes to go to Kapnapahar and at the tramway I met Heathcote. Stopped at the tea house where G. was waiting to complain but I had no time to listen as the outturn advice had to be sent. At 4 p.m. I left for Kapnapahar, where I found baby in splendid health. The doctor, as usual, didn't keep his promise and only called twice. Heathcote told me he had resigned and got a *locum tenens* in Assam from March 1st. It is a good thing he is going, as his ways are unsuitable.

Sat. Oct. 31st

After paying tickets none of the people made a move and on my speaking about it Gurudoyal sirdar shouted to them to make their complaint. Of course, I wouldn't stand for this and soon got a shift on, though it was touch and go for a minute whether they would obey. However, I told them I would listen to them at my own convenience and I sent G. to his house – Mia Babu is on holiday. After *chota* I rode round the garden and called the people by twos and three, and then the whole system was brought to light. Previously they were afraid to complain, not knowing me well, but were now driven to it. There were endless stories of fines for nothing (the KB pocketing them), of people taken by force to his house at night and hammered, and of women being insulted and threatened with assault. The people say he has caused absconding and if he is sacked many will return. He has been behaving like this for several years and stole more tickets previously, as Allanson seldom went to *gunti*. The way he is hated is extraordinary. The Khitmutghar also told me about him and Etwari, the punkah boy.

222

Sun. Nov. 1st

Weekly reports, etc., in the morning and in the afternoon went to Kapnapahar. The KB tried to pump me and offered to explain things, which I didn't require.

Tues. Nov. 3rd

Expected the doctor in the morning but, as usual, he didn't come till just as I was off to Kapnapahar, and as I wanted him to see the medical indent, it delayed me. I dined at K. and Agnes is to come home tomorrow.

Wed. Nov. 4th

Mia Babu is now back so after *chota* I wrote to the KB, giving him his dismissal. During the afternoon I went to the office and the KB asked to leave at once. He has taken it so quietly that he must have something up his sleeve. Rode to Kapnapahar, where I found Kennedy, had tea and escorted Agnes and baby back. It is nice to be all at home again.

Sat. Nov. 7th

Took over books from the KB and paid him up. He made a speech trying to prove himself an injured person and said he hoped I would take him back at some future time. The Heathcotes came to tea. H. wants to buy my nurseries on the Clevedon road.

Tues. Nov. 10th

Wired Lees if I could come to Surma to talk over estimates.

Wed. Nov. 11th

Heard from Lees I could come.

Thurs. Nov 12th

Left at 10 a.m. At the station met the doctor, also Etkins, who was meeting his brother-in-law, Hutchinson, just out from home. We talked for a time in the post office and I got a faint sort of fit and had to sit down. Sproull, who is staying with Lees, met me with the buggy but he had a riding pony as well, so he rode while I drove. Lees was waiting for me and thought I looked better. We started business at once and after dinner I made fair copies of the estimate and was tired by the time I got into bed.

Fri. Nov. 13th

Had tea about 4 a.m. and left at 5.30 a.m. The train was very late and I got tired of waiting in the cold. I had *chota* on its arrival and then changed to a carriage where I could lie down. I found all in order at Ruthna and had an easy afternoon to make up for an early start.

Sat. Nov. 14th

Outturn for fortnight 115 maunds, against ninety last year.

Sun. Mon. 15th

The Heathcotes dined with us and I had my dress things out for the first time in this bungalow.

Mon. Nov. 17th

Nothing to enter except the agents have sent for me to see a Calcutta doctor.

Tues. Nov. 18th

Intended to go to Clevedon but somehow didn't feel quite up to it. Agnes was also feeling a little seedy.

Sat. Nov. 21st

Got a wire from Lees, saying he would arrive tomorrow.

Sun. Nov. 22nd

Sent the pony for Lees but it returned alone. However, Lees suddenly arrived at 6 p.m., having come by the Assam mail and biked up. He seemed in a good temper.

Mon. Nov. 23rd

Lees and I went round and this time to the far end of Montri Joom, which he had never seen before. Some good *matti* for *bheel* soiling had been discovered there. During the afternoon we went to the office and tea house.

Tues. Nov. 24th

Breakfasted at Dhami. Lees biked and I rode at a slow canter. I had not ridden so far since I arrived here and felt stiff. Fergy has a nice bungalow and a first class fowl run. He goes in for breeding good stock. At breakfast the

225

talk was chiefly about the doctor and I was able to explain several matters. After we inspected the tea house, which is nicely arranged, we left at 5 p.m., and on reaching home I found Agnes lying down with a feverish headache.

Wed. Nov. 25th

Lees went to see the ghat road and *tillahs* at the back of the lines. He wishes me to extend here first in preference to the Clevedon road, as it will be cheaper. Agnes in bed all day with fever. I have decided to take her to Calcutta to see a doctor, as she is always getting seedy.

Fri. Nov. 27th

Went round the hoeing of the clearance, the work is not up to much. Rode to Kapnapahar for tea. Mrs Heathcote has promised to look after baby when we leave which will probably be Sunday.

Sun. Nov. 29th

Agnes took baby over early. Baby was carried and Agnes went in the pony rickshaw. I went to Kapnapahar for breakfast and shortly after went to the station. We got a carriage to ourselves. At Grimogal we went to the dining car and had quite a decent dinner afterwards, dozing in our own compartment till we reached Chandpur about 2 p.m. The train stops opposite the boat and we had a reserved cabin. The boat left at 4 p.m.

Mon. Nov. 30th

It was very cold on deck and we didn't reach Goalundo

till after breakfast. We got a place for ourselves in the train and at Sealdah a Grand Hotel man met us and fixed up everything. On our arrival at the hotel we had dinner in our own room. Robbia the Khitmugur accompanied us.

Tues. Dec. 1st

After *chota* I went to the office and saw Edwards. He gave me a letter to Major Bird, I.M.S., so I returned to the Grand Hotel and, taking Agnes, set off to Middleton Row. The Major, after questioning me, examined me under the X-rays, and it was very funny to look at my hand as if it were transparent. Then he found my spleen was enlarged and finally ordered me to a seaside place, Puri for choice, for a period of three to four months. He examined Agnes and said she only wanted a thorough change. At breakfast we were accosted by Major Herbert, who has come to meet his wife. He was looking very well and I was pleased to see him. Edwards dined with us and went afterwards to the theatre to see the Hugh Ward Company in *When Knights were Bold*. It was very amusing. The leading lady was Grace Palotta, whom we used think of as Mrs Kingston of East Dereham. The play began at 8.30 p.m., which is very late.

Wed. Dec. 2nd

Went to the office. They are cabling home about my *locum tenens*, so I have to wait another day. Got a dose of fever and was in bed all the rest of the day.

Thurs. Dec. 3rd

Went to the office but still no reply to the cable, so Edwards said I could leave. Did some shopping with Agnes and had some ices at Pelitis. After tea we went to the zoo and saw the rhino and a mithun, but the snakes were no good, as many had died and the rest were asleep. After dinner Bowery came over for a few minutes and then we sat with the Herberts.

Fri. Dec. 4th

Rose at 4.45 a.m. and left the hotel at 5.10 a.m. Nothing of interest on the journey. Arrived Chandpur at 7.30 p.m. and travelled on the slow train, leaving 9.30 p.m.

Sat. Dec. 5th

Reached Juri about 11 a.m. Here the coolies' ponies, etc. were assembled. Went straight to Kapnapahar and found baby in first class health. Returned to Ruthna and turned in at the tea house to see my new KB, who seems a quiet, capable man. Everything is in order and there are numerous old coolies returning.

Sun. Dec. 6th

Started getting arrears of work in the office cleared up, my reports to the 30th not having been sent off. Got a letter from McLeod, saying that Wedderspoon and his wife would arrive tomorrow.

Mon. Dec. 7th

Assam mail was late so the Wedderspoons did not arrive till nearly 8 a.m. He is elderly, cleanshaven, and much more like a padre or schoolmaster type than a planter. She is dark, rather small with blue eyes. They are much more like what we have been used to and conversation is a pleasure.

228

Wed. Dec. 9th

Went to Ellapore where the top dressing has been started at Montri Joom. At tea Mrs Harrison, the doctor and Sandeman called, the latter staying till 7 p.m.

Fri. Dec. 11th

Got my accounts sent off and things straightened up a bit.

Sun. Dec. 13th

Packed and gave over cash to Wedderspoon. Also, after breakfast, took him down to the office and arranged recruiting and programme of work.

Mon. Dec. 14th

Shut up all loose things and at 12 o'clock the mail arrived, telling of Father's* death on November 20th. His end was peaceful and not unexpected and now looking back, we as a family can thank God we had for a father one of the finest men He has ever put breath into; certainly the best man I have ever known.

At 2.30 p.m. we sent off our marl and followed ourselves at 3.15 p.m., myself on Rodney, then baby, carried in her basket, nurse in the coolie rickshaw and lastly Agnes in the pony rickshaw. We had a string of thirty coolies.

At the station baby caused much interest as she had her bottle on the platform. When the train came in Ferguson got out with another man I didn't know. Our compartment was reserved and we brought our dinner with us. The only servant we had was the Khitmagur. Lees got in at Itakhola

*Rev. H. Hetherington.

229

but we didn't see him till Chandpur, where he got on the mail, and we stayed in the train till morning.

Tues. Dec. 15th

Took our things to the waiting room and I made friends with the stationmaster, an Englishman, who took Agnes to his bungalow while I stayed with baby. They had *chota* there and he sent some to me. The boat came in about 10 a.m. and we reached Barisal about 5 p.m. I heard today that Aswini Babu, the organiser of the boycott, had been deported, which will do a great deal of good.

Wed. Dec. 16th

Woke up at a ghat which I think must be where the Chinese colony was in March 1906, but it seemed much more extensive now. The sunderbunds have been opened up to cultivation a great deal and we seemed to go quite a different route, but at 3 p.m. we went through a very narrow part, where it would have been impossible to have passed another boat. After this it was jungly on one side and cultivated on the other, almost till nightfall.

Thurs. Dec. 17th

All day we kept crossing broad pieces of water, which ran south as far as one could see, and I should think were really arms of the sea. By the evening we had passed Diamond Harbour.

Fri. Dec. 18th

When I got up we were at Kidderpore dock, landing tea,

and we didn't move on till 9 a.m. Also, when we approached Armenian Ghat we had to wait as Howrah Bridge was raised and we didn't get tied up till 1 p.m. After this there was trouble getting the luggage onto bullock carts and the drive through the streets was a nightmare, and the whole space was covered with bullock carts, drivers shouting, cabbies cursing, and bobbies making no regulation of the traffic. We reached the Grand Hotel safely and were quartered on the ground floor of the annex beyond the theatre, where there is a separate entrance. After a feed we took baby for a drive and called on Shirley but he had left for England.

Sat. Dec. 19th

Drove to bank to raise a loan but failed and got one in the office instead. Met Lees there and discussed the bungalow. We may get a *chung* after all. Returned to the hotel and met Percy Briscoe there. He has chucked the Empire and is to act for Arthur Moore this summer. Baby had her photo taken at 11.30 a.m. by Bourne and Shepherd, but she would not laugh properly, which was a pity.

I have engaged a bearer. He speaks English, which I don't like, and looks rather cheeky, but there was no time to look about much. Bowrey came with us to the station, which was a great help. The Bengal Nagpur Railway is a much broader gauge than the EBSR or the Assam Bengal and we had a very comfortable carriage. The guard was an Irishman and therefore very friendly. At Khargpur he had dinner fetched into our compartment, each plate covered with another, so as there were six courses and two of us, we had a pile of twenty four plates.

Sun. Dec. 20th

When we woke the train was running through level

231

country, rather sandy looking. At about 8 a.m. we reached Puri, which is a terminus. A man from the sanatorium was there to meet us and arranged gharries for us and the luggage, so we had no trouble,

The roads in Puri are good, mostly red brick dust, and trees are grown on either side. There are a number of fine houses belonging to natives between the station and the sanatorium and no sign of coolie huts. The sanatorium is about a mile and a quarter from the station and lies on the sand, perhaps 300 yards from the sea. There is no pretence of a compound and it gets every wind that blows.

We were met by the junior member, Mrs Bently, and given a room upstairs, which was small, uncarpeted, ill-furnished and dreary. A dressing room was attached of the same description. The doors were ill-fitting and the whole place a mass of draughts. A notice said the charge was Rs.8 for a single and Rs.14 for a double room on this floor and Rs.7 for a single and Rs.12 for a double room on the ground floor. Extra was charged for attendance at table if one's bearer does not wait. I found, to my annoyance, that my new man was a strict Hindoo and refused to. It seems to me fearfully expensive for the class of accommodation and we must make some bargain for a reduction.

We had *chota* of one fish course followed by toast and marmalade, which was not a big meal and left us hungry. We had breakfast at 12.30 p.m. which was a big meal with too many courses. After tea we went out in a gharry (a closed one, as there are no first gharries in Puri) and had dinner in our bedroom. Miss Forbes, the proprietor, is an energetic-mannered person with rather a bossy way and she and Mrs Bently wear nurse's costume.

Mon. Dec. 21st

Usual lack of hot water, and though the place swarms with mosquitoes there are no nets. There are no bells and everything is done on the cheap. There is not even a supply of bathroom lamps. Nurse got fever.

Wed. Dec. 22nd

Got a slight dose of fever. This place is the home of winds and draughts.

Fri. Dec. 25th

Christmas Day. Agnes has a fever, temperature of 100°. Lewis gave a tea party in the collectors' tent on the beach. Some new people called Dibbs have arrived. Agnes got up for Christmas dinner, which was quite good. One outsider, Captain McKilver, I.M.S., was present. Miss Forbes has her eye on him. Afterwards there was music in the so-called drawing room a tiny room with one or two chairs.

Mon. Dec. 28th

Agnes' fever still continues. Captain McKilver visited her.

Thurs. Dec. 31st

Nurse left by the 8.30 p.m. I could not afford to keep her here.

Fri. Jan. 1st, 1909

Today is cloudy and the air damper. My rheumatism is worse in consequence.

Sun. Jan. 3rd

Every day the same. *Chota* at 8.30 a.m., then baby has her bath while I read the paper; 10.30 a.m. my bath and afterwards write or loaf till breakfast for half an hour or so. Return and lie down till dinner or sometimes sing for a while.

Sat. Jan. 9th

Have discharged my Calcutta bearer and taken on a boy and an ayah, the boy is brother to the doctor's bearer.

Sun. Jan. 17th

My right ankle has swollen badly and I can't walk again. Things don't improve at all.

Sat. Jan. 24th

Mrs Bentley is ill with dysentery and the place more disorganised than ever.

Wed. Feb. 3rd

Some people called Andrews are sitting at our table. He is an American Presbyterian padre and is in charge of a school at Mussorie. We like them very much and he is to christen baby. They have travelled a great deal and he attended a 'cure' at Geneva for facial paralysis and advised me to try for it for my rheumatism. He only paid five francs per day at a hotel, which is very cheap.

Sun. Feb. 7th

Mr Andrews christened baby at 4.30 p.m. in the dining room, his wife, Mrs Bentley, Agnes and myself being present. The service commenced with prayer, then readings from the New Testament, followed by questions concerning faith, which we answered. Then was the sprinkling of water and baptismal words, followed by a paragraph from the prayer book: 'We receive this child . . .' Then the padre, contrary to the Presbyterian custom, signed the cross on her forehead; then another prayer, the blessing and all was over. Baby behaved very prettily and when the padre held her she laid her cheek against his in a most confiding way. He was quite moved and kissed her.

Wed. Feb. 10th

My knee is very swollen now and I am confined to my room.

Fri. Feb. 12th

We have now got a very good ayah. She is a Christian, but is not spoilt as so many of them are.

Sun. Feb. 14th

Took baby to be weighed at the gaol. She weighed 15lbs, while Agnes was 7st 9lbs and I was 8st 7lbs. Just the same as when I arrived. I hoped I had put on weight.

Sun. Feb. 21st

Doctor reports me unfit for work and recommends my going home.

Tues. Feb. 23rd

McLeods sanction leave and I think will help with my passage. Edwards has gone home.

Fri. Feb. 26th

The Lieutenant Governor, Sir Edward Baker, visited Puri. We drove round by the bazaar and everywhere was crowded while flags and triumphal arches were in evidence.

Sat. Mar. 6th

Agnes down again with fever.

Tues. Mar. 9th

Agnes got up and went for a drive in the open gharry drawn by two donkeys. We went by the broad road to Juggernath and had a fine view of the temple.

Tues. Mar. 16th

Have booked a cabin by the Anchor line *Algeria* sailing on April 16th.

Sat. Mar. 20th

Hear that *Algeria* carries no surgeon, so cancel berth.

Fri. Mar. 26th

Offered a cabin on the *City of Benares*, sailing on April 6th, which we accept.

Sat. Apr. 3rd

Did a lot of packing and after tea drove round to see the bed of spikes. We saw an occupant of one but he was not lying down, and the other was swung under a tree with a cloth over it. The evening was lovely and we stayed till after sunset.

Sun. Apr. 4th

Our last day, thank goodness. Spent the morning packing and after I took a rickshaw to the station to arrange a compartment. We left about 8.30 p.m. taking the ayah and bearer. Our carriage was comfortable and we managed to sleep.

10

RETURN HOME

Mon. Apr. 5th

Had *chota* at Khargpur and reached Calcutta at about 10
a.m. A man from KH & Co. was there; there was also a
durwan from the Grand. Agnes went with the latter
straight to the hotel, while I went first to Gladstone, Wylie
& Co. to arrange about going on board during the evening,
which concession was made, and then to McLeods. I saw
Bowrey and he told me things were against me at home,
as Russel thinks it should have been reported earlier by
both Lees and myself, and also the secretaries write that
they don't think that I shall be able to come out again.
This was rather a knockdown blow, but I consoled myself
with a long glass of iced Munich beer.

Arrived at the hotel at 1 p.m. and found we had splendid
rooms on the ground floor: bed, bath, sitting and dressing
rooms. We had a first class tiffin, a great treat after Puri.
At 5 p.m. we went for a drive to Bathgates, then round
by the river and maidan to the cathedral. At the verger's
house we met Dr Cogan and had a talk. Then sunset and
our last drive was very enjoyable. We dined in our room
and after dinner Bowrey came round and brought his wife.

Tues. Apr. 6th

Worked hard all the morning at KH & Co. Then to the

238

bank, from which I withdrew the silver boxes. Returned to lunch at 2 p.m. An hour later we settled up and at 3.30 p.m. we left. Agnes called in at Whiteways and got a wheelchair for baby, which, by working a spring, can be lowered and made into a rocking chair. We reached Kidderpore at 4.15 p.m. and though the boat was lying at No. 11 we had to go to No. 12 for the medical examination. The ship's doctor, the steward and some officers were there but the government official didn't arrive till 5 p.m. and his work was, of course, a farce.

On going aboard we found our cabin was small and very hot, but the durwan and bearer straightened it up and then I dismissed them, and also the ayah. This boat was originally built for cargo and there is very little deck space, but all the cabins, saloons, etc., are on the upper deck, so ports need never be shut. There is also a bridge deck with a few cabins and music room. We had quite a decent dinner and after Agnes turned in I stayed up till 11 p.m. to see the boat start hauling out of dock.

Wed. Apr. 7th

The passengers came aboard about 8 a.m. and at 9 a.m. we had our first general meal. Agnes and I were placed at the Captain's table, holding about twelve people, exclusive of the skipper and doctor. The people seem inclined to be friendly. There are two other babies but their mothers are travelling alone. The first person to speak to me was a Mr Taylor, F.R.G.S., elderly, travelling with his wife. He told me he had suffered with rheumatism and the best cure was at Ischia near Naples, as he had been cured there. At dinner some people dressed; some wore dark coats while others didn't change.

Fri. Apr. 9th

The voyage is quite cool and the only fault is the heat of the cabin. The food is good and the stewards obliging, while there is a library free of charge for all with quite good books in it.

Sun. Apr. 11th

Ceylon came into sight during the afternoon but the view was poor, as there was a good deal of cloud. Baby continues to give less trouble than we expected.

Mon. Apr. 12th

Got into Colombo harbour about 11 a.m. The heat was awful. I had hoped to go ashore in the afternoon for a drive to Mt Lavinia but found I could not get a rowing boat, which was a great disappointment to us.

Tues. Apr. 13th

A very hot morning. There were a number of people selling curios and Agnes bought an ebony elephant for Rs.8. We sailed about 2.30 p.m. and it got cooler immediately.

Wed. Apr. 14th

A few deck games have been started and Agnes has entered for deck billiards and quoits. Her partner is a Douars planter.

Thurs. Apr. 15th

There are quite a number of planters on board: Luard from Longai Valley, Florence from North Sylhet, Pringle from Balisera, Sloan from the Douars and a number of others. The army is represented by Colonel Maxwell from the North West Frontier.

Fri. Apr. 16th

The doctor examined me and is trying thirty grams per day of aspirin instead of six grains, as at Puri. He seems to know more than others who have fooled with me.

Sat. Apr. 17th

Agnes suffering from backache and headache and baby in consequence is troublesome. She has won four out of five games.

Mon. Apr. 19th

Passed Aden during the forenoon. It was terribly hot all day and night.

Wed. Apr. 21st

A strong breeze got up and it looked as if there was a sandstorm. The evening was quite cool.

Fri. Apr. 23rd

I left off white clothes and we had no punkah at night. It

has been most remarkably cool for the Red Sea.

Sat. Apr. 24th

Morning found us in the Gulf of Suez and the *Syria* was sighted on the horizon. A big Messagires boat passed us, looking very fine. We reached Suez about 7 p.m., when the doctor came aboard. I saw him in the bathroom, which saved going into the saloon. We left about 10 p.m.

Sun. Apr. 25th

The Canal was uninteresting and we reached Port Said about 4 p.m. We didn't go ashore but were pestered by the usual crowd of sellers. One man wanted Rs.24 for a bracelet but finally offered it with another for Rs3. We left about 7.30 p.m.

Thurs. Apr. 29th

Anchored in Malta harbour about 7 a.m. Agnes, baby and I went ashore at 8 a.m., accompanied by a Cook's man to help me. I was lifted in and out of the rowing boat and had no difficulty. On the landing side the town rises up very steeply and the buildings very solidly constructed. We got a carriage which went at quite a decent rate, considering the ascent. We visited the church of St John, which hasn't a very imposing outside but the inside was gorgeous with carving everywhere, while the roof was arched and painted overall. We should have liked to stay for a good while, but mass was to commence so we came away. Had *chota* at the Hotel Angleterre, beautifully cooked ham and eggs and the best coffee since Ruthna. From there we went back to the ship and Agnes looked in at the palace gardens for a few minutes en route. We reached the ship about 9.45 a.m.

242

and the bill for the Cook's man was 14s., made up of boat, 3s., carriage, two hours, 6s., guide, two hours, 5s.

We left about 10.45 a.m. and passed two German ships of war lying in the harbour. The Emperor is to arrive in a few days. The wind was bitterly cold and the sea showed signs of getting up.

Fri. Apr. 30th

The African coast was in sight off and on all day.

Sun. May 2nd

Passed the Straits of Gibraltar about 8 p.m. It was a moonlight night and the dark mass of the rock showed up finely with the powerful light at its foot, while another beacon could be seen on the African coast. The passage looked very narrow and once through it the lights of Gibraltar town appeared on the west side, making a very pretty sight.

Tues. May 4th

Passed Cape Finistère during the afternoon, so are now in the Bay. The boat rolled a bit but no one was sick.

Wed. May 5th

Very cold so stayed in the cabin all day and even there could hardly get warm.

Thurs. May 6th

Many people have colds. The last sweep was held and baby won the second prize, £1. 7s. 0d.

Fri. May 7th

When I got up we had just passed Dover and were very close in to the shore. The sea was choppy, green and cold-looking. At *chota* the *Mantua* passed us and I suppose Harry is on board. After finishing our meal we packed up and by 11.30 a.m. were off what we thought was Gravesend. At noon we were at Tilbury and heard to our annoyance that we were to be landed shortly in a tender. I engaged a King's man and a johnny was sent from the office to help me, so we had plenty of assistance. This was needed, as I had to go and interview the Customs people and climbing over trunks was hard work. Tips were as follows: cabin steward, 7s. 6d., bath boy, 2s. 6d., table steward, 5s., stewardess 10s., on the whole very cheap.

About 1.30 p.m. we left on the tender. A regular gale was blowing and it was icy cold. We were met at Fenchurch Street and walked from there to Mark Lane, where the luggage had been taken on a hand cart. Arrived at Agnes' mother's house in Hampstead for tea. We had a good night, sleeping on decent beds at last.

Sat. May 8th

Rita took me to the City. We walked to the LN & WR and NLR station at West End Lane, where the trains run every quarter of an hour and booked to Broad Street, which is the terminus. Here we engaged a motor taxicab and went to Lime Street. I had never been in a taxi before and enjoyed both the pace and skill of the chauffeur. Russel was not in the office so went on to King and Co. in Cornhill to see about the luggage. Got back about 2 p.m. I was dead tired and spent the afternoon on the sofa.

Mon. May 10th

Took a taxi to see a specialist in Harley Street, Dr Hale White. After examination he decided that I should try Dr Bernard Scott's baths in Bournemouth. He also said I must have my blood examined by Dr Eyre of Guy's Hospital.

Wed. May 12th

Called at the office and saw Russel, who asked me questions about my rheumatism and said the board would make an allowance and that I ought to look for a job at home. He was quite decent but didn't seem anxious for me to go back to India.

A Note Made in the Diary, dated 1931

I did not return to Assam but never forgot my first view of the snows and ever afterwards they had an extraordinary fascination for me. They were not visible during the rains and at their best mid-October to end of February; also the light was best from midday. I used to mount to the top of the tea house ladder every morning, or nearly so, for a view. They must have been over fifty miles distant and rose above the lower ranges.

APPENDIX: WORK ROTA, MARCH 1903

Daily Kamjari 24-3-08

	Men	Women	Children
Establishment	29	1	2
Paniwallas	9		
Cowherds	7		3
Line chowkidars	5		
Night chowkidars	5		
Garden chowkidars	3		
Jongles chowkidars	2		
Syces	5		
Malis	2		1
Hoeing No. 20	117		3
Hoeing No. 23	2	66	3
Cutting edges	3		
Weeding No. 2	7	70	18
With bricklayers & getting bricks	30	1	34
Railing and fencing	11		
Spreading leaf			4
Carpenters	12		
Carters	11		
Dakwalas	3		
Blacksmiths	4		
Dhulies Paniwalas	2		
Sweepers	2		
Nursing sick	2		
Gathering manure	3		
Cutting grass	18		
Tica carters	7		
With tica carters	7		
Making baskets	3		

	Men	Women	Children
Pound chowkidar	2		
Coal chowkidar	1		
Catching crickets			6
Cutting posts	10		
Sewing panda	1		
Total	326	138	74
Recruits	6	1	
Sick	23	20	2
Personnel in lines	32	94	19
Total	387	253	95

Absent
Moslaia
Mongol VII
Gopigiroo
Sukoo V
Moondi II

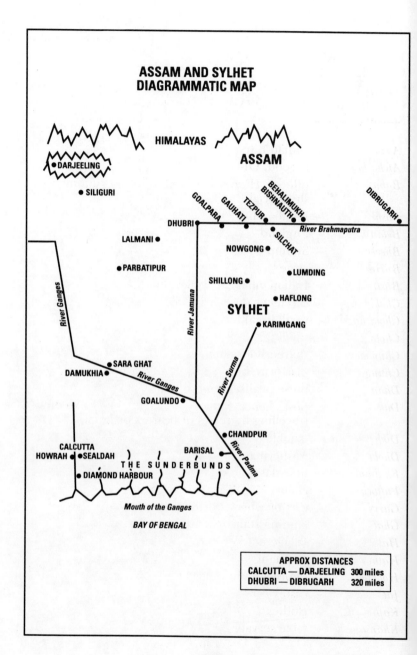

ASSAM AND SYLHET
DIAGRAMMATIC MAP

HIMALAYAS

ASSAM

• DARJEELING

• SILIGURI

GOALPARA GAUHATI TEZPUR BEHALIMUKH BISHNAUTH DIBRUGARH

DHUBRI • *River Brahmaputra*

• LALMANI NOWGONG • SILCHAT

• PARBATIPUR SHILLONG • • LUMDING

River Ganges *River Jamuna* • HAFLONG

SYLHET

• KARIMGANG

• SARA GHAT *River Surma*

DAMUKHIA • *River Ganges*

GOALUNDO •

• CHANDPUR

CALCUTTA BARISAL *River Padma*

HOWRAH • • SEALDAH T H E   S U N D E R B U N D S

• DIAMOND HARBOUR

Mouth of the Ganges

BAY OF BENGAL

APPROX DISTANCES	
CALCUTTA — DARJEELING	300 miles
DHUBRI — DIBRUGARH	320 miles

248

GLOSSARY

Ayah	– Indian nurse
Atcha hai	– is good, OK
Babu	– title of Hindus, an Indian clerk or one with a superficial English education
Bashi	– Bamboo hut
Belait	– England
Bheel	– Swamp
Burra	– large
Busti	– Indian village or collection of huts
Chokidar	– caretaker or watchman
Chota hazri	– small breakfast
Chota	– small
Chummery	– shared household
Chung	– platform, house on raised platform
Dhan	– pulse pealike plant
Dak	– post, hence mail a letter or parcel. Method of travelling by relays of bearers or horses
Dikh	– trouble
Dhobi	– washerman
Ek dum	– first-class
Futtock	– chains
Garry	– cart or wheeled vehicle for hire
Ghat	– landing stair or place
Hat	– village market
Hathi	– elephant
Hoolah	– depression in the ground
Jat	– caste
Kapre	– clothes
Khansamah	– table servant

Kirani	–	headman or head clerk
Kutcha	–	crude, unfinished
Mah	–	raft
Mali	–	gardener
Mahout	–	elephant driver
Marl	–	luggage
Matti	–	clay
Mistrie	–	mechanic or craftsman
Mugger	–	crocodile
Muggra	–	bad tempered
Maund	–	eighty pounds weight
Napet	–	barber hairdresser
Nautch	–	dance
Paragon	–	tea-drying stove
Peg	–	whisky and soda
Pujah	–	hindu holiday
Punkah	–	cloth suspended from the ceiling and pulled by a *punkah wallah* or latterly electrically driven fan
Syce	–	groom
Salaam karoo	–	make a salutation or call
Simpkin	–	champagne
Tamasha	–	entertainment, show, fuss
Ticca	–	pay hire
Tank	–	artificial lake
Tiffin	–	lunch
Tillah	–	low hill
Tum tum	–	dogcart

LETTER OF APPOINTMENT

THE IMPERIAL TEA COMPANY, LIMITED.

TELEGRAMS.
"CLURICAUNE, LONDON."
TELEPHONE N° 430 AVENUE.

10. & 11 Lime Street,
London, 2nd August 1900
E.C.

Frank A Hetherington Esq,

    10 Sea Cliff Terrace

        Sea Cliff Road

            Bangor

                Co Down.

Dear Sir,

    We have pleasure in informing you that you have been appointed as Assistant at this Company's Tarajulie Division for three years on the terms and conditions mentioned in the enclosed Agreement The Company will pay all your expenses to the Garden including messing and hotel bills while in Calcutta. Your salary will beginnfrom the date of your joining your appointment.

    We have secured you a berth in the s/s "Himalaya" connecting with the s/s "Oriental" at Aden sailing from Marseilles on Thursday 16th Inst. The train which runs in connection with this steamer leaves London, (Charing Cross) @ 9 a.m. on Wednesday 15th To get this comfortably you will require to be in London on the 14th when we shall be glad to see you and when we shall hand you Rail and Passage tickets, as also sufficient funds to meet your requirements until you reach Calcutta.

    With regard to your baggage, you should send all your heavy luggage to the Royal Albert Docks, London as early as possible. The accompanying circular will give you the necessary information

    We enclose for your signature and return 2 copies of the Agreement which we propose giving you. A copy will be sent you immediately after the next Directors Meeting.

        Yours faithfully

        Stewart McLeod

           Secretaries.

251